Handbook of Dielectric and Thermal Properties of Materials at Microwave Frequencies

For a listing of recent titles in the
Artech House Microwave Library,
turn to the back of this book.

Handbook of Dielectric and Thermal Properties of Materials at Microwave Frequencies

Vyacheslav V. Komarov

ARTECH

HOUSE

BOSTON | LONDON
artechhouse.com

Library of Congress Cataloging-in-Publication Data
A catalog record for this book is available from the U.S. Library of Congress.

British Library Cataloguing in Publication Data
A catalog record for this book is available from the British Library.

ISBN-13: 978-1-60807-529-4

Cover design by Vicki Kane

© 2012 Artech House

10 9 8 7 6 5 4 3 2 1

And ye shall know the truth,
and the truth shall make you free
St. John, 8, 32

Contents

3

Foodstuff and Agricultural Products   29

4

Biological Tissues   69

5

Fibrous Materials   81

6

Polymers, Resins, and Plastics   91

7

Ceramics   101

Preface

The application of microwave energy for thermal processing of different materials and substances is a rapidly growing trend of modern science and engineering. Successful design and optimization of microwave heating applicators is impossible without theoretical studies of the physical processes of electromagnetic wave interaction with dielectric lossy media. Mathematical modeling and experimental measurements are the main tools for investigation of such processes. Many analytical and numerical models are used today for predicting temperature distribution in the interaction domain. Most of these models are based on the coupled system of differential equations for electromagnetic and thermal fields with corresponding boundary conditions. Scientists and engineers who are engaged in modeling microwave heating processes know how important the information about dielectric and thermal properties of microwave exposed samples is. This information can be obtained either experimentally by direct measurements or found in the literature, where such data is quite odd. First, it is very difficult to find both the dielectric and thermal properties of one material. Second, dielectric properties of materials are usually represented at room temperature while the temperature-dependent properties are the most interesting. Third, the necessary information on the complex dielectric permittivity of media is not always published at so-called industrial, scientific, and medical (ISM) frequencies, which are 430 MHz, 915 MHz, 2.45 GHz, 5.8 GHz, and 24.125 GHz allocated for microwave heating purposes.

The general objective of this book was to compile a survey of dielectric and thermal properties of microwavable materials as a function of temperature (or moisture) at ISM bands or at frequencies close to these bands.

Data represented in this book was collected for seven groups: food materials (Chapter 3), biological tissues (Chapter 4), fibrous materials (Chapter 5), polymers, resins and plastics (Chapter 6), ceramics (Chapter 7), soils and minerals (Chapter 8), and pure and composite chemical substances (Chapter 9). Some of dielectric media can belong to different groups; for example, many foods are biological tissues (meat, eggs, fish, etc.), or the water that we drink every day is made up of chemical substances.

One more peculiarity of this book is that temperature and moisture dependencies of physical parameters are given in the form of analytical equations, mainly polynomial expressions, which sometimes agree with experimental data very well and sometimes only approximately. In this case, the reader is recommended to look at the reference where initial dependencies in the form of figures are represented. But in most cases such approximation can be useful when the reader plans to employ some commercial packages—for example, COMSOL for numerical modeling microwave heating processes.

This book is intended mainly for engineers in the industry and researchers in universities who are engaged in applied electromagnetics, microwave power engineering, numerical modeling, and computer-aided design (CAD) of different microwave devices, food and agricultural engineering, physical chemistry, material science, and so forth. I also hope that this book will be interesting to students with corresponding backgrounds in physics.

The following organizations are acknowledged for the financial support of this research from 1997 to 2011: Swiss Academy of Engineering Sciences, Switzerland; Swedish Institute, Sweden; Industrial Microwave Modeling Group, Worchester Polytechnic Institute, Worcester, MA; US Army Natick Soldier Center, Natick, MA; German Academic Exchange Service (DAAD), Germany; Ministry of High Education and Science of Russian Federation.

I would be grateful for any remarks or notification of errors from the readers of this book.

1

Mathematical Modeling of Microwave Heating Processes

Presently, along with traditional methods of heating diverse materials, microwave energy is widely utilized for the thermal treatment of food, ceramics, wood, polymers, chemical substances, biological tissues, and so on. Deep penetration of microwaves in dielectric media improves uniformity and intensifies the heating process. The variety of microwave heating devices such as kitchen ovens, industrial plants, medical applicators, and laboratory setups is very large.

The development of applied electrodynamics, the theory of numerical modeling, and computational software and hardware has led to the appearance of different mathematical models for simulating electromagnetic and thermal fields in microwave heating systems. The most well-known among them is the coupled electromagnetic heat transfer problem, which takes into account the influence of temperature on distribution of microwave power sources in an interaction domain. Few modifications of this problem for solid samples have been considered in [1–8]. But as it has been shown in [9–11], formulation of the coupled problem for liquid media is more complicated and includes electromagnetic heat transfer problem as a particular case.

Propagation of electromagnetic waves in linear, isotropic, and homogeneous solid or liquid media is described by Maxwell's equations:

$$rot\dot{H} = j\omega\dot{\varepsilon}(T)\varepsilon_0\dot{E} + \dot{J}_{cm} \tag{1.1}$$

$$rot\dot{E} = -j\omega\mu_0\dot{H} \tag{1.2}$$

$$div\varepsilon_0\varepsilon'(T)\dot{E} = \rho_{cm} \tag{1.3}$$

$$div\dot{H} = 0 \tag{1.4}$$

where $\dot{E}$ and $\dot{H}$ are the complex amplitudes of the E- and H-field in space and $E(\tau) = \mathrm{Re}(\dot{E}e^{j\omega\tau})$, $H(\tau) = \mathrm{Re}(\dot{H}e^{j\omega\tau})$, ρ_{cm} is the specific density of source charge; $\dot{J}_{cm}$ is the source current density; ω is the angular frequency, ε_0, μ_0 are the dielectric and magnetic constants, $\dot{\varepsilon} = \varepsilon' - j\varepsilon''$ is the complex dielectric permittivity of material, $j = \sqrt{-1}$, T is the temperature; τ is the time. The magnetic permeability $\mu' = 1$ and magnetic loss factor $\mu'' = 0$ for dielectric media.

Equation (1.3) can be rewritten as

$$div\varepsilon_0\varepsilon'(T)\dot{E} = \varepsilon'(T)div\dot{E} + \left[\dot{E}, grad\varepsilon'(T)\right] = \frac{\dot{\rho}_{cm}}{\varepsilon_0} \tag{1.5}$$

$$div\dot{E} + \left[\frac{\dot{E}}{\varepsilon'(T)}, grad\varepsilon'(T)\right] = \frac{\dot{\rho}_{cm}}{\varepsilon_0\varepsilon'(T)} \tag{1.6}$$

And combining (1.1) and (1.2), we obtain

$$rotrot\dot{E} = \omega^2\varepsilon_0\mu_0\dot{\varepsilon}(T)\dot{E} - j\omega\mu_0\dot{J}_{cm} \tag{1.7}$$

$$graddiv\dot{E} - \nabla^2\dot{E} = \omega^2\varepsilon_0\mu_0\dot{\varepsilon}(T)\dot{E} - j\omega\mu_0\dot{J}_{cm} \tag{1.8}$$

$$\nabla^2 \dot{E} + \omega^2 \varepsilon_0 \mu_0 \dot{\varepsilon}(T) \dot{E} + grad\left[\frac{\dot{E}}{\varepsilon'(T)}, grad\varepsilon'(T) \right] = \frac{1}{\varepsilon_0} grad\left(\frac{\dot{\rho}_{cm}}{\varepsilon'(T)} \right) + j\omega\mu_0 \dot{J}_{cm} \qquad (1.9)$$

Introducing the wave number of free space $k_0^2 = \omega^2 \varepsilon_0 \mu_0$ and using the equation of continuity

$$div \dot{J}_{cm} = -j\omega\dot{\rho}_{cm} \qquad (1.10)$$

we can rewrite (1.9) as

$$\nabla^2 \dot{E} + k_0^2 \dot{\varepsilon}(T) \dot{E} + grad\left[\frac{\dot{E}}{\dot{\varepsilon}(T)}, grad\varepsilon'(T) \right] = j\omega\mu_0 \dot{J}_{cm} - \frac{1}{j\omega\varepsilon_0} grad\left(\frac{div \dot{J}_{cm}}{\varepsilon'(T)} \right) \qquad (1.11)$$

When $\dot{J}_{cm} = 0$, (1.11) is transformed in the well-known [1, 3] homogeneous Helmholtz equation for temperature dependent dielectric media

$$\nabla^2 \dot{E} + k_0^2 \dot{\varepsilon}(T) \dot{E} + grad\left[\frac{\dot{E}}{\dot{\varepsilon}(T)}, grad\varepsilon'(T) \right] = 0 \qquad (1.12)$$

Solutions of (1.11) must satisfy Neumann and Dirichlet boundary conditions on metal walls, the condition of continuity of tangential components of electromagnetic (EM) field on media interfaces, and scattering boundary conditions in lossy media.

The absorbed microwave energy is computed from the electromagnetic field distribution

$$q_v(\overline{r}, \tau) = 0.5\omega\varepsilon_0 \varepsilon''(T) \dot{E}^2 \qquad (1.13)$$

where ε'' is the loss factor.

In some simple cases, as for example thin dielectric rod in rectangular waveguide, power density is defined as:

$$q_v = P_0 \left(1 - \sum_{n=1}^{N} |S_{n1}|^2 \right) V_d^{-1} \qquad (1.14)$$

where V_d is the sample volume, P_0 is the input power, and S_{n1} is the absolute value of n-parameter of the scattering matrix.

The system of momentum, continuity, and energy conservation differential equations for temperature and flow velocity fields in Boussinesq's approximation and with several assumptions done in [9–11] is utilized for simulation of microwave heating of non-Newtonian liquids

$$\rho_t\left(\partial W_x/\partial \tau + \left(\tilde{W}\nabla\right)W_x\right) = \mu_t\nabla^2 W_x - \partial p/\partial x \qquad (1.15)$$

$$\rho_t\left(\partial W_y/\partial \tau + \left(\tilde{W}\nabla\right)W_y\right) = \mu_t\nabla^2 W_y - \partial p/\partial y \qquad (1.16)$$

$$\rho_t\left(\partial W_z/\partial \tau + \left(\tilde{W}\nabla\right)W_z\right) = \mu_t\nabla^2 W_z - \partial p/\partial z - g\rho_t\beta_t\left(T - T_a\right) \qquad (1.17)$$

$$div\tilde{W} = 0 \qquad (1.18)$$

$$\rho_t C_t\left[\frac{\partial T}{\partial \tau} + \left(\tilde{W}\nabla\right)T\right] = \lambda_t\nabla^2 T + q_v\left(\overline{r},\tau\right) \qquad (1.19)$$

where μ_t is the dynamic viscosity, g is the acceleration due to the gravity, β_t is the volume thermal expansion coefficient, p is the pressure, W_x, W_y, W_z are the components of the velocity vector $\tilde{W}$ in Cartesian coordinates, T_a is the ambient temperature.

Solutions of Navier-Stokes's equations (1.15) to (1.17) must satisfy Marangoni's and zero slip boundary conditions together with initial condition $T = T_0$, $\tau = 0$. Equation (1.19) must be also completed by the initial condition and boundary conditions for temperature. The Newton's boundary condition is used the most frequently:

$$-\lambda_t\left(T\right)\frac{\partial T}{\partial n} = \alpha_t\left(T_{int} - T\right) \qquad (1.20)$$

where T_{int} is the interface temperature, ρ_t, C_t, λ_t are the density, heat capacity, and thermal conductivity of media respectively, α_t is the heat transfer coefficient, depending on many parameters including sample sizes and surrounding media determined from

Nusselt's number $Nu = f(Pr, Gr)$, where Pr, Gr are the Prandtl and Grashof numbers, respectively, for liquid media. Heat convection intensity depends on the Grashof number.

Temperature distribution in an interaction zone for solid media is defined from solution of heat transfer equation

$$\rho_t(T)C_t(T)\frac{\partial T}{\partial \tau} = \lambda_t(T)\nabla^2 T + q_v(\bar{r}, \tau) \tag{1.21}$$

which is a particular case of (1.19) for $\tilde{W} = 0$. The heat transfer coefficient for solid samples is calculated from Bio's number.

The generalized coupled electromagnetic, heat transfer, and natural convection problem for microwave heating of liquids is formed by (1.11), (1.13), and (1.15) to (1.19) with corresponding boundary conditions. The coupled EM-heat transfer problem usually utilized in the modeling microwave heating processes is based on (1.11), (1.13), (1.20), and (1.21).

In many cases not only the aggregate state of irradiated sample but also the operating temperatures, possible chemical or biological reactions, peculiarities of particular technology realization, and other factors have an influence on mathematical model formulation [12].

For example, sometimes thermodiffusion and pressure inside the sample along with internal moisture transfer must be taken into account. In this case, the so-called coupled electromagnetic, heat, and mass transfer problem [13, 14] will additionally include differential equations for moisture and pressure. A successful solution to this problem requires information about such parameters as the specific moisture capacity of vapor phase, the coefficient of phase transformation air-vapor, the thermal diffusitivity, the diffusion coefficients describing the changes of mass and pressure, the specific heat of evaporation, and so forth.

Sometimes interaction of electromagnetic waves with lossy media is accompanied by variations of aggregate state (melting, thawing, etc.) of a specimen [15, 16]. Then enthalpy formulation of the energy balance equation with a superficial mushy region around the melting point [16] and Stefans's boundary conditions for moving front [15] can be successfully utilized for predicting temperature patterns in an interaction domain.

When microwave radiation is used for the heating of biological tissues, the heat transfer equation (1.21) is transformed in the Pennes's bioheat equation [17, 18] that contains the specific power density caused by biochemical processes in human body, the volumetric flow of blood, the blood heat capacity, and some other parameters.

An additional term in the heat transfer equation also appears when the temperature of heated material exceeds 300°C [19]. In this case, the power density of thermal radiation will be determined by the Stefan-Boltzmann constant, the coefficient of radiation energy directivity, the emission coefficient, and the size of the surface.

A similar situation takes place for exothermic chemical reactions [20]. Here, variations of temperature will depend on not only microwave power density but also on exothermic reaction itself and the following parameters: the gas constant, the thermal energy produced by one mole reactant, the initial concentration of the solution; these factors must be taken into account.

Coupled electromagnetic, heat transfer, and thermal stress problems must be solved when internal thermal stress is observed in microwave-exposed solid materials. The system of coupled differential equations for this problem is formed by many thermal and mechanical parameters—in particular, the vector of volumetric forcers, the tensor of deformations, the variation of the specific volume, the coefficient of specific volume variations in the area of elastic deformations, the shift module, the function of relative shift of the material, the components of mechanical stresses tensor, and the derivatives along space coordinates [21].

Variations of dielectric and thermal properties of materials versus temperature impact the uniformity of microwave heating. One of the serious reasons for uneven distribution of the thermal field in lossy media is known as the thermal runaway phenomenon. A given effect is characterized by the uncontrolled rise of temperature in the local area inside a sample. If microwave power is not changed, thermal runaway can be explained by a rapid increasing of the loss factor $\varepsilon''(T)$. Such dependence is observed for some ceramic and polymer materials [13]. According to (1.13), the higher ε'' more energy is absorbed. And if dependence $\varepsilon''(T)$ is strong, a microwave-exposed sample will be overheated and damaged.

Thermal runaway also takes place if microwave power is slightly changed during the heating process. This phenomenon is described by the so-called S-shaped curve [22, 23]. There are several theoretical models explaining the unstable behavior of the thermal field. One of them is based on the resonance nature of thermal runaway when the phase constant and the attenuation coefficient affect power dissipation in different materials [24, 25]. But some others prove the key role of $\varepsilon''(T)$ [26] or the thermal conductivity [27] in this phenomenon. The analytical 1-D electrothermal model described in [28] has shown that not only the thermal conductivity but geometrical sizes and temperature of a sample can influence microwave heating uniformity.

Different 1-D [1, 3, 4, 28], 2-D [11, 15], and 3-D [2, 5, 10] coupled models are successfully utilized today for numerical simulation of EM waves interaction with lossy media in microwave applicators using mainly numerical techniques: the finite element method (FEM) [2–5, 11, 15], finite difference time domain (FDTD) method [7, 10, 19], finite volume time domain (FVTD) method [29], transmission line matrix method [8], and others. Examples of a successful numerical solution of the coupled problem for microwave heating processes by means of commercial software are represented in [30–32].

Linear or nonlinear character of the coupled problem is defined by functions $\varepsilon'(T)$, $\varepsilon''(T)$. Nonlinear problem takes place when dependence of complex dielectric permittivity $\dot{\varepsilon}(T) = \varepsilon'(T) - j\varepsilon''(T)$ is strong. Weak dependence $\dot{\varepsilon}(T)$ leads to the linear or quasi-linear coupled problem. In both cases duration of thermal processes is much longer than EM ones and it is necessary to use either two time scales [4] or so-called adiabatic approximation according to which physical properties of material are not changed during several periods of EM field oscillation [29].

For linear coupled problem $q_v(\overline{r}, \tau) = const$ and for nonlinear $q_v(\overline{r}, \tau, T) = var$. That is solution of nonlinear microwave heating problem requires application of iterative algorithm with subsequent solution of EM and thermal differential equations for small time steps during which we can neglect variations of $\dot{\varepsilon}(T)$. In this case, the coupled problem solution is more accurate but computational time will essentially increase.

Nonlinear coupled problem formulation is necessary in most simulations of microwave heating systems because of quite strong dependencies of $\dot{\varepsilon}(T)$ for many media. In particular, nonlinear analysis is very important for food and ceramic materials for which thermal runaway phenomenon is observed.

However, in many practical cases variations of dielectric properties of some microwave-exposed materials are not strong and it is difficult to predict if it is necessary to solve the nonlinear problem or if it is enough to restrict analysis by linear problem and hence save computer memory.

Comparison of dielectric and thermal parameters of different media shows that temperature dependencies of thermal conductivity, heat capacity, and density for them are much weaker than $\dot{\varepsilon}(T)$ in most cases when phase transitions are not considered. Among materials that demonstrate relatively weak variations $\dot{\varepsilon}(T)$ at industrial, scientific, and medical (ISM) frequencies one can mention resins [13], low moisture content wood [33], liquid polymers [34], some minerals and soils [35], human blood [36], and silver oxide [37].

Independently on the behavior of functions $\varepsilon'(T)$, $\varepsilon''(T)$ during microwave heating, we can introduce a dimensionless parameter that characterizes variation rate of complex dielectric permittivity

$$\dot{\eta} = \eta' - j\eta'' \qquad (1.22)$$

where $\eta' = \varepsilon'_{min}/\varepsilon'_{max}$, $\eta'' = \varepsilon''_{min}/\varepsilon''_{max}$. Here $\varepsilon'_{min} - j\varepsilon''_{min}$ and $\varepsilon'_{max} - j\varepsilon''_{max}$ are the minimum and maximum values of dielectric permittivity and the loss factor of a heated specimen in operating temperature range at fixed frequency.

So, for all dielectric materials

$$0 < \{\eta', \eta''\} \leq 1 \qquad (1.23)$$

and when $\{\eta', \eta''\} \rightarrow 0$ we deal with a nonlinear coupled problem and when $\{\eta', \eta''\} \rightarrow 1$ with linear one. Then in the real practice of mathematical modeling of microwave heating processes it is important to know the left limit of range (1.23) that allows obtaining an adequate solution of the coupled problem in linear approximation without significant loss of accuracy.

Sometimes it is necessary to find the depth at which the microwave power decreases to e^{-1}, where e is the Euler's constant, of its value at the surface of lossy media. The given parameter, known as penetration depth, can be calculated [13] using data about ε' and ε'':

$$D_p = \frac{\lambda}{2\sqrt{2}\pi\sqrt{\varepsilon'}\left[\sqrt{1+\left(\dfrac{\varepsilon''}{\varepsilon'}\right)^2}-1\right]^{0.5}} \tag{1.24}$$

Here λ is the operating wavelength of microwave source. The distance at which the incident microwave power is decayed to one-tenth of its initial value:

$$Z_p = \frac{1.15129}{\alpha} \tag{1.25}$$

$$\alpha = 1.48 \cdot 10^{-8} f \sqrt{\varepsilon'\left(\sqrt{1+\left(\dfrac{\varepsilon''}{\varepsilon'}\right)^2}-1\right)} \tag{1.26}$$

where α is the attenuation coefficient.

References

[1] Jolly, P., and I. Turner, "Non-linear Field Solutions of One-Dimensional Microwave Heating," *Int. J. Microwave Power and Electromagnetic Energy*, Vol. 25, No. 1, 1990, pp. 4–15.

[2] Dibben, D. C., and A. C. Metaxas, "Finite Element Time Domain Analysis of Multimode Applicators Using Edge Elements," *Int. J. Microwave Power and Electromagnetic Energy*, Vol. 29, No. 4, 1994, pp. 242–251.

[3] Ayappa, K. G., H. T. Davis, E. A. Davis, and J. Gordon, "Analysis of Microwave Heating of Materials With Temperature-Dependent Properties," *AIChE Journal*, Vol. 37, 1991, pp. 313–322.

[4] Alpert, Y., and E. Jerby, "Coupled Thermal-Electromagnetic Model for Microwave Heating of Temperature-Dependent Dielectric Media," *IEEE Transactions on Plasma Science*, Vol. 27, No. 2, 1999, pp. 555–562.

[5]   Sekkak, A., L. Pichon, and A. Razek, "3D FEM Magneto-Thermal Analysis in Microwave Ovens," *IEEE Transactions on Magnetics*, Vol. 30, No. 5, 1994, p. 3347–3350.

[6]   Soriano, V., C. Devece and E. De los Reyes, "A Finite Element and Finite Difference Formulation for Microwave Heating Laminar Material," *Int. J. Microwave Power and Electromagnetic Energy*, Vol. 33, 1998, pp. 67–76.

[7]   Ma L., D. L. Paul, N. Pothecary, C. Railton, J. Bows, L. Barratt, J. Mullin and D. Simons, "Experimental Validation of a Combined Electromagnetic and Thermal FDTD Model of a Microwave Heating Process," *IEEE Transactions on Microwave Theory and Techniques*, Vol. 43, No. 11, 1995, pp. 2565–2572.

[8]   Flockhart, C., V. Trenkic and C. Christopoulos, "The Simulation of Coupled Electromagnetic and Thermal Problem in Microwave Heating," *Proceedings of the 2nd Conference on Computation in Electromagnetics*, Nottingham, UK, 1994, pp. 267–270.

[9]   Ayappa, K. G., S. Brandon, J. J. Derby, H. T. Davis and E. A. Davis, "Microwave Driven Convection in a Square Cavity," *AIChE Journal*, Vol. 40, 1994, pp. 1268–1272.

[10] Zhang, Q., T. H. Jackson and A. Ungan, "Numerical Modeling of Microwave Induced Natural Convection," *Int. J. Heat and Mass Transfer*, Vol. 43, 2000, pp. 2141–2154.

[11] Ratanadecho, P., K. Aoki and M. Akahori, "A Numerical and Experimental Investigation of the Modeling of Microwave Heating for Liquid Using a Rectangular Wave Guide (Effects of Natural Convection and Dielectric Properties)," *Applied Mathematics Modeling*, Vol. 26, 2002, pp. 449–472.

[12] Komarov, V. V., "Formulations of the Coupled Mathematical Models of Microwave Heating Processes," *Int. J. Applied Electromagnetics and Mechanics*, Vol. 36, No. 4, 2011, pp. 309–316.

[13] Metaxas, A.C. and R. J. Meredith, *Industrial Microwave Heating*, London: Peter Peregrinus, 1983.

[14] Rogov, I. A. and S. V. Nekrutman, *Ultrahigh Frequency Heating of Food Products*, Moscow, Russia: Agropromizdat, 1986 (in Russian).

[15] Ratanadecho, P., K. Aoki, and M. Akahori, "The Characteristics of Microwave Melting of frozen Packed Beds Using a Rectangular Waveguide," *IEEE Transactions on Microwave Theory and Techniques*, Vol. 50, No. 6, 2002, pp. 1495–1502.

[16] Basak, T., and K. G. Ayappa, "Influence of Internal Convection During Microwave Thawing of Cylinders," *AIChE Journal*, Vol. 47, No. 4, 2001, pp. 835–850.

[17] Pennes, H. H., "Analysis of Tissue and Arterial Blood Temperature in the Resting Human Forearm," *Journal of Applied Physics*, Vol. 1, 1948, pp. 93–122.

[18] Hardie, D., A. J. Sangster, and N. J. Cronin, "Coupled Field Analysis of Heat Flow in the Near Field of a Microwave Applicator for Tumor Ablation," *Electromagnetic Biology and Medicine*, Vol. 25, 2006, pp. 29–43.

[19] Haala, J. and W. Wiesbeck, "Modeling Microwave and Hybrid Heating Processes Including Heat Radiation Effects," *IEEE Transactions on Microwave Theory and Techniques*, Vol. 50, No. 5, 2002, pp. 1346–1354.

[20] Huang, K. M., Z. Lin, and X. Q. Yang, "Numerical Simulation of Microwave Heating on Chemical Reaction in Dilute Solution," *Progress in Electromagnetics Research Online*, Vol. 49, 2004, pp. 273–289.

[21] Arkhangelskiy, Yu. S. and S. V. Trigorliy, *UHF Electro Thermal Radiation Setups*, Saratov: Saratov State Technical University Issue, 2000 (in Russian).

[22] Wu, X., *Experimental and Theoretical Study of Microwave Heating of Thermal Runaway Materials*, Ph.D. Dissertation, Virginia Polytechnic Institute, 2002.

[23] Liu, B., T. R. Marchant, I. W. Turner, and V. Vegh, "A Comparison of Semi-Analytical and Numerical Solutions for the Microwave Heating of a Lossy Material in a Three-Dimensional Waveguide," *Microwave and Radio Frequency Applications*, The American Ceramic Society, 2003, pp. 17–27.

[24] Vriezinga, C. A., "Thermal Runaway in microwave Heated Isothermal Slabs, Cylinders, and Spheres," *Journal of Applied Physics*, Vol. 83, No. 1, 1998, pp. 438–442.

[25] Vriezinga, C. A., "Thermal Profiles and Thermal Runaway in Microwave Heated Slabs," *Journal of Applied Physics*, Vol. 87, No. 7, 1999, pp. 3774–3779.

[26] Spotz, M. S., D. J. Skamser, and D. L. Johnson, "Thermal Stability of Ceramic Materials in Microwave Heating," *Journal of the American Ceramic Society*, Vol. 78, No. 4, 1995, pp. 1041–1048.

[27] Parris, P. E., and V. M. Kenkre, "Thermal Runaway in Ceramics Arising from the Temperature Dependence of the Thermal Conductivity," *Physica Status Solid. B-Basic Research*, Vol. 200, No. 1, 1997, pp. 39–47.

[28] Paulson, M., L. Feher, and M. Thumm, "Parameters Optimization Modeling Using Stationary 1D-Electrothermal Model to Improve Temperature Homogeneity," *Int. J. Microwave Power and Electromagnetic Energy*, Vol. 39, No. 3 & 4, 2004, pp. 141–151.

[29] Zhao, H., and I. W. Turner, "The Use of a Coupled Computational Model for Studying the Microwave Heating of Wood," *Applied Mathematical Modeling*, Vol. 24, 2000, pp. 183–197.

[30] Sabliov, C. M., D. A. Salvi, and D. Boldor, "High Frequency Electromagnetism, Heat Transfer and Fluid Flow Coupling in ANSYS Multiphysics," *Int. J. Microwave Power and Electromagnetic Energy*, Vol. 41, No. 4, 2007, pp. 5–17.

[31] Kopyt, P., and M. Celuch, "Coupled Electromagnetic-Thermodynamic Simulations of Microwave Heating Problems Using the FDTD Algorithm," *Int. J. Microwave Power and Electromagnetic Energy*, Vol. 41, No. 4, 2007, pp. 18–29.

[32] Komarov, V. V., Thumm M., Feher L., and Akhtar J., "Heating Patterns of Microwave Exposed Liquid Polymers," *Proceedings of the Global Congress on Microwave Energy Application*, 2008, Otsu, Japan, pp. 845–848.

[33] Torgovnikov, G. I., *Dielectric Properties Of Wood and Wood Based Materials*, Berlin, Germany: Springer Verlag, 1993.

[34] Akhtar, M. J., L. E. Feher, and M. Thumm, "A Novel Approach for Measurement of Temperature Dependent Dielectric Properties of Polymer Resins at 2.45 GHz," *Proceedings of the Global Congress on Microwave Energy Application*, 2008, Otsu, Japan, pp. 529–532.

[35] Salsman, J. B., "Measurement of Dielectric Properties in the Frequency Range of 300 MHz to 3 GHz as a Function of Temperature and Density," *Proceedings of the Symposium Microwaves: Theory and Application in Materials Processing*, Cincinnati, USA, 1991, pp. 203–214.

[36] Stuchly, M. A., S. S. Stuchly, "Dielectric Properties of Biological Substances—Tabulated," *Int. J. Microwave Power and Electromagnetic Energy*, Vol. 15, No. 1, 1980, pp. 19–26.

[37] Atwater, J. E., "Complex Dielectric Permittivity of the Ag_2O-Ag_2CO_3 System at Microwave Frequencies and Temperatures Between 22°C and 189°C," *Applied Physics*, A, 2002, 75, pp. 555–558.

2

Dielectric and Thermal Properties of Microwaveable Materials: Parameters, Measuring Techniques, and Some Theoretical Aspects

2.1 Complex Dielectric Permittivity

Theoretical fundamentals and mechanisms of electromagnetic waves interaction with lossy dielectric media are well described in the literature [1–7]. Most microwave technologies (aside from some specific trends such as sintering of metal powders) [8] are intended for thermal processing of diverse dielectric materials such as food, wood, ceramics, polymers, and so forth. All these materials absorb microwave energy due to the dipole polarization phenomenon when their polar molecules try to align themselves to the external electromagnetic field. Since the field reverses with high frequency, energy produced when going from an ordered form to a random form is converted into heat.

The polarization ability of materials is characterized by the so-called relative dielectric permittivity:

$$\varepsilon' = 1 + N\frac{\alpha_p}{\varepsilon_0} \tag{2.1}$$

where α_p is the polarizability of the medium, and N is the particles concentration.

There are several mechanisms of polarization including electronic, ion (atomic), dipolar, and interfacial (Maxwell-Wagner). All these mechanisms also contribute to one more important parameter: dielectric loss factor, which is determined in the microwave range as:

$$\varepsilon'' = \varepsilon_d'' + \frac{\sigma}{\varepsilon_0\omega} \tag{2.2}$$

where ε_d'' is the dielectric loss due to the dipole polarization, and σ is the ionic conductivity.

Electronic and atomic types of polarization are observed mainly at infrared and optical frequencies; that is, they can be neglected in (2.1) and (2.2).

So, the dielectric properties of materials are described by the complex dielectric permittivity:

$$\dot{\varepsilon} = \varepsilon' - j\varepsilon'' \tag{2.3}$$

The ratio of the imaginary part of (2.3) to the real one is called the tangent of loss angle:

$$tg\delta_e = \frac{\varepsilon''}{\varepsilon'} \tag{2.4}$$

Both parameters ε' and ε'' are affected by frequency, temperature, moisture, and pressure [6]. Taking into account that most microwave applicators operate at normal pressure and fixed frequencies, $\dot{\varepsilon}$ depends on two dominant factors: temperature and moisture. Influence of these factors on $\dot{\varepsilon}$ for various media are well-studied [1, 2, 6, 7].

The majority of dielectrics subjected to microwave processing are composite systems consisting of two or more components forming spatially disordered statistical mixtures. The physical properties of composite media are most frequently studied experimentally by measuring their dielectric parameters. However, in some cases, such measurements prove to be impossible due to the absence of the required equipment or the material specimen. In this case, the physical properties of materials may be simulated using the theory of dielectric mixtures (DMs). The key element in implementation of this approach consists of the initial information about the properties of all the components in the composite media under consideration, and in particular of water whose polar molecules interact with the microwave field in the moist-containing media.

Currently, the DM theory is widely used in engineering electrodynamics for calculation effective complex permittivity of bioelectromagnetic phantoms, shielding elements, radar absorbing materials, and so forth. Inhomogeneous media described by this theory are usually classified into three classes [3].

The first class includes dispersive media in a continuous phase that are the mixtures where one component is considered as a uniform medium with a permittivity ε_1 in which particles of another uniform medium ε_2 are distributed in a regular or random way. In the general case, these particles may have any shape (spherical, elliptical, disclike, needlelike, etc.).

The second class (continuous media in a continuous medium) contains inhomogeneous media consisting of "layers" of two materials that are distributed (or interwoven) in such way that in each of the media every two points are connected by a continuous path.

The third class contains dispersive media in dispersive media consisting of separated particles of materials 1 and 2, which do not depend on other parts of the same material, with the possible inclusion of air pores.

Many DM formulas are represented and analyzed in books [9, 10]. Some examples of the implementation of the DM theory application to microwave heating processes and materials are presented in [11–15]. In some cases, the formulas of the DM theory were tested. Study [11] was devoted to the applicability of the formulas by Wiener, Fricke-Madgett, Böttcher, Lichtennecker, and Landau-Lifshitz for the description of the permittivity of mixtures (ε_m) of aqueous emulsions and suspensions at a frequency of 2.45 GHz. It was found that the values of ε_m for emulsions of water-oil type are best described by the Fricke-Madgett formula and ε_m for emulsions of lime-water type, by the Lichtennecker formula. The values of ε_m were measured in [11] at T = 20°C.

In study [12], the DM formulas were derived for ε_m of two- and three-component mixtures, which may be adapted for spherical, ellipsoidal, cylindrical, and needlelike inclusions. The formula that takes into account ellipsoidal inclusions of water yielded approximate agreement with the experimental data for ε_m of pine-wood at 2.45 GHz and at room temperature.

Several formulas of the DM theory for assessment of ε_m of moist wood at different temperatures were tested in [15]. Seven DM models that yielded the minimal error were tested. Calculations were made for three frequencies using published experimental data and the values of the permittivity for the tap water at the frequency 915 MHz.

The most reliable source of information about dielectric properties of pure and composite materials of any aggregate state is direct experimental measurement. There are many techniques developed for such measurements [16–18].

2.1.1 Open-Ended Coaxial Probe Method

The open-ended coaxial probe (OCP) method is currently one of the most popular techniques for measuring of complex dielectric permittivity of many materials. Nondestructive, broadband (RF and microwave ranges), and high-temperature (up to 1,200°C) measurements can be carried out with this method using commercially available instrumentation. Its well-developed theory makes it possible to obtain sufficiently accurate results both for medium-loss and high-loss media [19–21].

A typical OCP system consists of an automatic network analyzer (ANA) with a calibration kit, custom built test cell, programmable circulator, coaxial cable, and personal computer connected to the ANA through a special bus (Figure 2.1). The material under study is placed in a steel pressure-proof test cell. The probe is

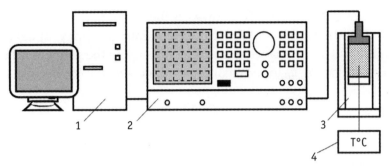

Figure 2.1 Schematic diagram of experimental setup: computer (1), ANA (2), measurement cell with the sample (3), and thermometer (4).

kept in a close contact with the sample during the measurements via a steel spring and piston. A thin rigid stainless thermocouple probe passes onto the center of the sample to measure sample temperature.

The sensing element of an OCP system is an open-ended cylindrical coaxial line that is excited by transverse electromagnetic (TEM) wave. Parameters (amplitude and phase) of incident and reflected signals are detected by the ANA. The complex dielectric permittivity is determined according to the reflected coefficient ($\Gamma = \Gamma' - j\Gamma''$) as follows [22]:

$$\varepsilon' = \left(A_e f\right)^{-1} \left\{ \frac{-2\Gamma''}{\left(1+\Gamma'\right)^2 + \Gamma''^2} \right\}; \quad \varepsilon'' = \left(A_e f\right)^{-1} \left\{ \frac{1-\Gamma'^2 - \Gamma''^2}{\left(1+\Gamma'\right)^2 + \Gamma''^2} \right\} \tag{2.5}$$

where A_e is the empirical coefficient dependent on characteristic impedance of the probe and sample size. In order to eliminate the influence of reflections caused by transmission line discontinuities, a calibration procedure is utilized. The EM characteristics of measurement system are analyzed using three standard terminations (open, short, and 50 Ω). Then any material with well-known dielectric properties such as deionized water, for example, is tested. The actual reflection coefficient differs from reflection coefficient measured using ANA (Γ_m) [20]:

$$\Gamma = \frac{\Gamma_m - a_{11}}{a_{22}\left(\Gamma_m - a_{11}\right) + a_{12}} \tag{2.6}$$

where a_{11} is the directivity error, a_{12} is the frequency response error, and a_{22} is the source match error. Taking into account propagation constant (γ) and distance from the connector to the probe head (z), we can calculate a_{ij} in terms of S-parameters of the connector:

$$a_{11} = S_{11}; \ a_{12} = S_{12}S_{21}e^{-2\gamma z}; \ a_{22} = S_{22}e^{-2\gamma z} \qquad (2.7)$$

In the inverse coaxial probe model, it is assumed that a sample has a semi-infinite size.

A few additional conditions must be satisfied to avoid measurement error in the OCP method:

- Minimize thermal expansion of both conductors of the coaxial line at high temperatures;
- Intimate contact between the probe and the sample: a liquid sample may flush the probe and the surface roughness of solid sample should be less than 0.5 μm [23];
- Minimize disturbance caused by temperature, vibration, or any other external factors after calibration and during the measurement.

The OCP method is very well suited for liquids or soft solid samples. It is accurate, fast, and broadband (from 0.2 to up to 20 GHz). The measurement requires little sample preparation. A major disadvantage of this method is that it is not suitable for measuring materials with low dielectric property (plastics, oils, etc.).

2.1.2   Transmission Line Method

The transmission line method (TLM) belongs to a large group of nonresonant methods of measuring complex dielectric permittivity of different materials in a microwave range [1, 24]. There exist several modifications to this method including the free-space technique [25], open-circuit network method (see Section 2.1.1), and short-circuited network method. Usually three main types of transmission lines are used as the measurement cell in TLM: rectangular waveguide, coaxial line, and microstrip line.

An analyzed sample is placed near the short-circuited end of a transmission line [7]. The dielectric properties of the sample are determined using the following expressions:

$$\varepsilon' = \left(\frac{\lambda}{2\pi d}\right)^2 \left(x^2 - y^2\right) + \left(\frac{\lambda}{\lambda_{qc}}\right)^2 ; \; \varepsilon'' = -\left(\frac{\lambda}{2\pi d}\right)^2 2xy \qquad (2.8)$$

where λ is the free-space wavelength, λ_{qc} is the quasi-cutoff wavelength, d is the sample thickness; $x = Re(Z_{in})$, and $y = Im(Z_{in})$; Z_{in} is the input impedance of the short-circuited line.

$$Z_{in} = \frac{K_t^2 + tg^2\left(2\pi/\lambda_a l\right)}{K_t\left[1 + tg^2\left(2\pi/\lambda_a l\right) + j\left(1 - K_t^2\right)tg\left(2\pi/\lambda_a l\right)\right]} \qquad (2.9)$$

where l is the distance between the dielectric surface and the first minimum of the standing wave, λ_a is the wavelength in unloaded part of transmission line, and K_t is the traveling wave coefficient which is calculated when $K_t \geq 0.4$ as:

$$K_t = \sqrt{\frac{E_{min}}{E_{max}}} \qquad (2.10)$$

and when $K_t < 0.4$ as:

$$K_t = \frac{\sin\left(2\pi\Delta x/\lambda_a\right)}{\sqrt{\left(\left(E/E_{min}\right) - 1\right)\sin\left(2\pi\Delta x/\lambda_a\right)}} \qquad (2.11)$$

where E_{min} and E_{max} are the minimum and maximum values of the electric field amplitude and $2\Delta x$ is the distance between two points along the transmission line on both sides of minimum where measured data are equal and determined from: $E = m^2 E_{min}$ ($2 < m < 10$ is the empirical coefficient found from the calibration procedure).

A schematic diagram of a typical experimental setup realizing the transmission line method is represented in Figure 2.2.

Dielectric permittivity of lossy media may be also successfully measured employing a two-port coaxial cell with a sample placed in the middle of transmission line, so that the TEM wave could propagate from the input port to the output port. Impedance

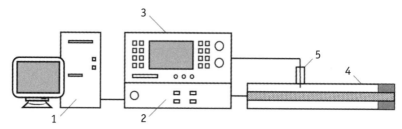

Figure 2.2 Microwave system for measuring dielectric properties of materials using the transmission line method: computer (1), microwave oscillator (2), detector (3), short-circuited piece of coaxial slotted line with specimen (4), and carriage with probe (5).

changes and propagation characteristics (S-parameters) of TEM wave measured by means of ANA in an empty and partially loaded transmission line leading to the determination of the dielectric properties of lossy material. The basic principles of this technique are given in [26].

In general, a TLM measurement system is more expensive for the same range of frequency than the open-ended coaxial probe system, and the measurements are more difficult and time-consuming. The method described above gives good accuracy for high lossy materials but has rigid requirements on sample shape and sizes. In particular, the sample shape needs to precisely fit the cross section of the transmission line. In some cases, in order to increase accuracy it is necessary to measure several samples of various thicknesses. Despite these drawbacks, TLM is still widely used in microwave measuring engineering due to its simplicity. Using of coaxial line as a basic unit of measurement cell makes this method sufficiently broadband. The accuracy of this method is generally between that of the OCP method and that of the resonance cavity method.

2.1.3 Resonant Cavity Method

Resonant cavity methods are also widely utilized in measuring complex dielectric permittivity of lossy materials. The most popular resonant cavity method is the perturbation method (PM), which is based on a comparative analysis of certain EM characteristics between empty and partially loaded rectangular or cylindrical resonance cavity [27–29] (Figure 2.3).

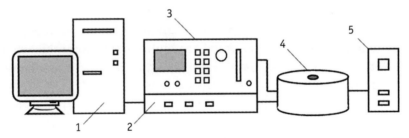

Figure 2.3 Schematic diagram of experimental setup: computer (1), microwave oscillator (2), detector (3), cavity resonator with a sample (4), and heating unit (5).

According to PM theory [28, 29], dielectric permittivity and losses of a sample under study are determined as following:

$$\varepsilon' = 1 + A^{-1}\left(\frac{V_c}{V_s}\right)\left(\frac{\Delta f}{f_0}\right); \quad \varepsilon'' = B^{-1}\left(\frac{V_c}{V_s}\right)\left(\frac{1}{Q_1} - \frac{1}{Q_0}\right) \tag{2.12}$$

where f_0 and Q_0 are the resonance frequency and Q-factor of the empty cavity, f_1 and Q_1 are the resonance frequency and Q-factor of the cavity with a sample, V_c is the cavity volume, V_s is the sample volume; $\Delta f = f_0 - f_1$. A and B are the coefficients that depend on several parameters: shape, sizes, and location of the sample in the cavity and configuration and excited operating mode of the cavity. In some cases, A and B may be found analytically for a lossy sheet material placed in a rectangular cavity with operating mode TE_{103} [27] or they may be determined empirically with calibration of the experimental setup. Equation (2.12) is valid when three main assumptions are satisfied [28]: the dielectric sample does not disturb the general distribution of EM field in the cavity; metallic wall losses do not influence the resulting losses in the cavity; Q_0 and Q_1 are measured at the same frequency. Appropriate location of the sample is also a very important factor, which affects the accuracy of the measurement. Preliminary numerical modeling of the microwave setup with lossy dielectric material inside the cavity may be a useful approach for determining an optimum sample position in this case [27]. Sometimes, measurement errors are possible when there are air gaps between the specimen and the conducting parts of the metallic resonator.

There are also some restrictions in using conventional resonance PM for measuring the dielectric loss tangent of low-loss media. If conduction losses in cavity walls are higher (or comparable) than dielectric losses of specimen, the resonator Q-factor may change and one will not obtain the correct values of the ε''. In this case, application of hybrid high-order modes called whispering-gallery modes [30] or the special calibration procedure of Q-factor characterization as a function of frequency [28] can help to eliminate this drawback of this method.

PM is more accurate than the waveguiding methods. It is particularly suited for medium-loss and low-loss materials and substances. Precise shaped small-sized samples are usually used with this technique. But PM provides dielectric properties measurements only at a fixed frequency. Commercial systems from Hewlett-Packard are more expensive than the open-end coaxial probe system.

2.2   Thermal Conductivity

Thermal conductivity (λ_t, [W/(m·K)]) is a very important parameter responsible for estimation of molecular transfer of heat in continuous media caused by temperature gradient. It is determined from the Fourier equation:

$$q = -\lambda_t grad T \tag{2.13}$$

where q is the heat flux density. Different units utilized for λ_t definition in different countries are given in Table 2.1 [31].

Table 2.1
Thermal Conductivity Units

Parameter	Equivalent Parameter, λ_t [W/(m·K)]
1 cal$_{IT}$/(cm·c·°C)	418.68
1 cal$_{th}$/(cm·c· °C)	418.4
1 Btu$_{IT}$/(ft·h·°F)	1.73073
1 Btu$_{IT}$·in/(ft^2·h·°F)	0.144228
1 Btu$_{th}$/(ft·h·°F)	1.72958
1 Btu$_{th}$·in/(ft^2·h·°F)	0.144131
1 CHU/(ft·h·°F)	3.1152

Here, $1\ cal_{IT} = 4.1868\ J$ is the international calorie, $1\ cal_{th} = 4.184\ J$ is the thermo chemical calorie, $1\ Btu_{IT} = 1055.05\ J$ is the British thermal unit, and $1\ Btu_{th} = 1054.35\ J$; $1\ CHU = 1899.1\ J$.

The given parameter is affected by temperature, pressure, substance aggregate state, purity, and other factors. The thermal conductivity of many liquids is increased when the temperature rises, except water and glycerin. In the temperature range $-50 < T°C < 50$ temperature dependence $\lambda_t(T)$ for liquids can be approximately calculated using expression [31]:

$$\lambda_t(T) = \lambda_t(0°C)\left[1 + 0.01T\left(\frac{\sqrt{T_b}}{23.5} - 1\right)\right]$$

(2.14)

where T_b is the boiling temperature of aqueous substance; $\lambda_t(0°C)$ is the thermal conductivity at $T = 0°C$, and also:

$$\lambda_t(0°C) = 0.009 \cdot N_a^{-0.25}\sqrt{T_b\rho_t C_p}$$

(2.15)

Here N_a is the number of atoms in liquid molecule, ρ_t is the density, and C_p is the specific heat capacity at constant pressure. For most liquid substances $0.1 \le \lambda_t, W/(m \cdot K) \le 1$.

Thermal conductivity of solid media is defined by its structure: porous materials have lower λ_t than continuous ones, because for air $\lambda_t = 0.025\ W/(m \cdot K)$. Then it has been found [32] that for denser solids thermal conductivity is greater. Moisture also causes an increase of the given parameter.

2.3   Heat Capacity, Density, and Viscosity

Heat capacity characterizes the ability of media to accumulate heat by increasing its temperature. For liquid substances, two parameters—specific heat capacity (C_t, $[J/(kg \cdot K)]$ and volumetric heat capacity (C_v, $[J/(m^3 \cdot K)]$—are applied, and are also linked as [31]:

$$C_t = C_v + \beta_t\frac{TV}{k_t}$$

(2.16)

where β_t is the volume thermal expansion coefficient, V is the substance volume, and k_t is the isothermal compression. For solid materials, $C_t \approx C_v$ and always $C_t > C_v$.

We can also define heat capacity employing one more well-known equation:

$$Q = MC_t\left(T - T_0\right) \tag{2.17}$$

where Q is the amount of heat transmitted to the media, M is the mass, and T_0 is the initial temperature.

Density is the ratio $\rho_t = M/V$ [kg/m^3]. Parameter $a_t = \lambda_t/(C_t \cdot \rho_t)$ [m^2/s] is called thermal diffusivity.

Dynamic viscosity (μ_t, [Pa·s]) is one more important parameter that must be known for analysis of processes of microwave heating of liquids. It is determined from Newton's friction law:

$$F_x = -\mu_t \frac{\partial W_x}{\partial y} \tag{2.18}$$

where F_x is the internal friction force directed perpendicular to the gradient of velocity of aqueous media. In some handbooks, one can find another parameter: kinematical viscosity $v_t = \mu_t/\rho_t$ [m^2/s]. Dynamic viscosity almost always does not depend on pressure but strongly depends on temperature.

Experimental methods of λ_t determination are based on the solution of the heat transfer equation. So-called steady-state methods are used for measuring λ_t at a fixed temperature. Different transient techniques—the transient plane source (TMS) method, modified TPS method, transient line source method, thermoreflectance method, Searle's bar method, and Lee's disc method—are utilized for investigation of λ_t as a function of temperature. These methods demonstrate good accuracy and flexibility in measuring λ_t of solid, liquid, powdered, and thin-film materials. There is a large variety of experimental tools for these purposes, including special calorimetrical devices and needle-type sensors [33, 34].

Calorimeters can be used also for experimental studies of the heat capacity. There are various modifications of these instruments: adiabatic, bomb, isothermal titration, constant-pressure, differential scanning, calvet-type, reaction, and so forth [35]. The measuring technique implemented in these calorimeters is based on (2.17), which is very well known in academic teaching. The same is true about the measuring of density and viscosity.

References

[1] Metaxas, A. C., and R. J. Meredith, *Industrial Microwave Heating*, London: Peter Peregrinus, 1983.

[2] Von Hippel, A., *Dielectric Materials and Applications*, Cambridge, MA: MIT Press, 1954.

[3] Roussy, G., and J. A. Pearce, *Foundations and Industrial Applications of Microwaves and Radio Frequency Fields*, New York: Wiley, 1995.

[4] Thuery, J., *Microwaves: Industrial, Scientific and Medical Applications*, Norwood, MA: Artech House, 1992.

[5] Okress, E. C., *Microwave Power Engineering*, New York: Academic Press, 1968.

[6] Komarov V. V., S. Wang, and J. Tang, "Permittivity and Measurement," *Wiley Encyclopedia of RF and Microwave Engineering*, Vol. 4, Hoboken, NJ: Wiley and Sons, 2005, pp. 3694–3711.

[7] Rogov, I. A., and S. V. Nekrytman, *Microwave Heating of Food Products*, Moscow: Agropromizdat, 1986 (in Russian).

[8] Roy, R., D. K. Agrawal, and J. P. Cheng, et al., "Full Sintering of Powdered Metals Using Microwaves," *Nature*, Vol. 399, No. 17, 1999, pp. 668–670.

[9] Neelakanta, P. S., *Handbook of Electromagnetic Materials: Monolithic and Composite Versions of Their Application*, New York: CRC Press, 1995.

[10] Shivola, A. H., *Electromagnetic Mixing Formulas and Applications*, London: IEE Books, Electromagnetic Wave Series, Vol. 47, 1999.

[11] Erle, U., M. Regier, C. Persch, et al., "Dielectric Properties of Emulsions and Suspensions: Mixture Equations and Measurement Comparisons," *International Journal of Microwave Power and Electromagnetic Energy*, Vol. 35, No. 3, 2000, pp. 185–190.

[12] Tinga, W. R., and W. A. G. Voss, "Generalized Approach to Multiphase Dielectric Mixture Theory," *Journal of Applied Physics*, Vol. 44, No. 9, 1973, pp. 3897–3902.

[13] Shutko, A. M., and E. M. Reutov, "Mixture Formulas Applied in Estimation of Dielectric and Radiative Characteristics of Soils and Grounds at Microwave Frequencies," *IEEE Transactions on Geoscience and Remote Sensing*, Vol. 20, No. 1, 1982, pp. 29–32.

[14] Fratticcioli, E., M. Dionigi, and R. Sorrentino, "A New Permittivity Model for the Microwave Moisture Measurement of Wet Sand," *Proceedings of the 33rd European Microwave Conference*, Munich, Germany, 2003, pp. 539–542.

[15]   Komarov, V. V., "Testing of Models Used in the Theory of Mixtures and Applied to Calculations of the Permittivity of Moist Wood Tissue in the Microwave Range," *Journal of Communications Technology and Electronics*, Vol. 52, No. 11, 2007, pp. 1245–1251.

[16]   Afsar, M. N., J. R. Birch, and R. N. Clarke, "The Measurement of the Properties of Materials," *IEEE Proceedings*, Vol. 74, 1986, pp. 183–199.

[17]   Bartnikas, R., "Dielectric Measurement," *Wiley Encyclopedia of Electrical and Electronics Engineering Online*, New York: Wiley, 1999.

[18]   Krupka, J., and R. G. Geyer, "Loss-Angle Measurement," *Wiley Encyclopedia of Electrical and Electronics Engineering Online*, New York: Wiley, 1999.

[19]   Stuchly, M. A., and S. S. Stuchly, "Coaxial Line Reflection Methods for Measuring Dielectric Properties of Biological Substances at Radio and Microwave Frequencies—A Review," *IEEE Transactions on Instrumentation and Measurement*, Vol. 29, 1980, pp. 176–183.

[20]   Blackham, D. V., and R. D. Pollard, "An Improved Technique for Permittivity Measurement Using a Coaxial Probe," *IEEE Transactions on Instrumentation and Measurement*, Vol. 46, 1997, pp. 1093–1099.

[21]   Otto, G. P., and W. C. Chew, "Improved Calibration of a Large Open-Ended Coaxial Probe for Dielectric Measurements," *IEEE Transactions on Instrumentation and Measurement*, Vol. 40, 1991, pp. 742–746.

[22]   de los Santos, J., D. Garsia, and J. A. Eiras, "Dielectric Characterization of Materials at Microwave Frequency Range," *Materials Research*, Vol. 6, 2003, pp. 97–101.

[23]   Arai, M., J. G. P. Binner, and T. E. Cross, "Estimating Errors Due to Sample Surface Roughness in Microwave Complex Permittivity Measurements Obtained Using a Coaxial Probe," *Electronics Letters*, Vol. 31, 1995, pp. 115–117.

[24]   Torgovnikov, G. I., *Dielectric Properties of Wood and Wood Based Materials*, Berlin, Germany: Srpinger-Verlag Publishing, 1993.

[25]   Ghodgaonkar, D. K., V. V. Varadan, and V. K. Varadan, "A Free Space Method for Measurement of Dielectric Constants and Loss Tangents at Microwave Frequencies," *IEEE Transactions on Instrumentation and Measurement*, Vol. 37, 1989, pp. 789–793.

[26]   Courtney, C. C., "Time-Domain Measurement of the Electromagnetic Properties of Materials," *IEEE Transactions on Microwave Theory and Techniques*, Vol. 46, 1998, pp. 517–522.

[27]   Komarov, V. V., and V. V. Yakovlev, "Modeling Control Over Determination of Dielectric Properties by Perturbation Technique," *Microwave and Optical Technology Letters*, Vol. 39, No. 6, 2003., pp. 443–446.

[28]   Chen L., C. K. Ong, and B. T. Tan, "Amendment of Cavity Perturbation Method for Permittivity Measurement of Extremely Low-Loss Dielectrics," *IEEE Transactions on Instrumentation and Measurement*, Vol. 48, 1999, pp. 1031–1037.

[29] Kraszewski, A. W., and S. O. Nelson, "Observation on Resonant Cavity Perturbation by Dielectric Objects," *IEEE Transactions on Microwave Theory and Techniques*, Vol. 40, 1992, pp. 151–155.

[30] Krupka, J., D. Cros, M. Auburg, and P. Guillon, "Study of Whispering Gallery Modes in Anisotropic Single-Crystal Dielectric Resonators," *IEEE Transactions on Microwave Theory and Techniques*, Vol. 42, 1994, pp. 56–61.

[31] Grigoriev, I. S., and E. Z. Meylihov (eds.), *Physical Constants: Handbook*, Moscow: Energoatomizdat, 1991 (in Russian).

[32] Lukanin, V. N. (ed.), *Thermal Engineering: Textbook for Universities*, Moscow: High School, 2002 (in Russian).

[33] Platunov, E. S. (ed.), *Thermophysical Measurements and Instruments*, Leningrad: Mashinostroenie, 1986 (in Russian).

[34] Tritt, T. M. (ed.), *Thermal Conductivity: Theory, Properties, and Applications*, Berlin: Springer, 2004.

[35] Höhne G., W. Hemminger, and H. J. Flammersheim, *Differential Scanning Calorimetry*, Berlin, Heidelberg: Springer-Verlag, 1996.

3

Foodstuff and Agricultural Products

General aspects and practical examples of using microwaves for thermal processing different food materials are well described in many publications including [1–5]. As it is known, the physical properties of food and agricultural products depend on moisture, salts, fats, proteins, and the content of various chemical substances. Temperature is another factor, which impacts on the complex dielectric permittivity (CDP), density, heat capacity, thermal conductivity, and viscosity of foods. Values of all these parameters can vary in wide ranges because of essential heterogeneity of food and agricultural products. As well, different groups of foodstuff demonstrate different physical properties, as shown below.

3.1 Meat

Meat is one of the most well-studied foods. For example, CDP values of meat with various moisture and fat content at ISM frequencies are given in Tables 3.1 and 3.2.

Table 3.1
Dielectric Properties of Different Meat Samples at $T = 20°C$

Meat	W%	Fat Content%	433 MHz		915 MHz		2,450 MHz	
			ε'	ε''	ε'	ε''	ε'	ε''
Beef	75.4	2.2	53.7	33.5	51.7	27.8	48.7	17.6
Pork	76.5	2.4	54.6	34.4	52.5	28.4	49.5	18.1
Mutton	76.6	1.5	53.9	33.6	51.6	28	48.9	18
Hen	74.6	0.9	52.6	37.6	49.4	30.1	46.2	20
Duck	74.3	1.2	52.1	35.3	47.8	30.1	46	20.2
Goose	74.2	1.8	53.9	34.5	48.4	29.7	47.6	19.2

From: [3].

Table 3.2
Dielectric Properties of Beef (Shoulder Part) at $T = 20°$

Moisture%	433 MHz		915 MHz		2,450 MHz	
	ε'	ε''	ε'	ε''	ε'	ε''
30	10.5	7	8.7	5.2	7.5	3.5
40	24.5	26.5	20.6	21.8	16.1	12
45	31.5	28.6	30	25.1	25.5	15.3
55	44	31	43.1	26.3	36	16.6
70	53	32.3	50.5	27	46.6	17.1
75	53.7	33	51.7	27.4	48.5	17.6

From: [3].

This data corresponds to room temperature, but in microwave tempering and defrosting technologies it is more important to know the CDP of meat at low temperatures (see Table 3.3).

Table 3.3
Dielectric Properties of Frozen Meat at $T = -20°C$

Meat sample	300 MHz		915 MHz		2,450 MHz	
	ε'	ε''	ε'	ε''	ε'	ε''
Beef	5.4	0.7	4.8	0.54	4.4	0.51
Pork	5.5	1.14	4.4	0.63	4.0	0.56
Chicken	5.3	1.31	4.6	0.87	4.0	0.51
Turkey	5.3	1.21	4.5	0.73	4.1	0.61

From: [3].

Temperature is a very important factor that influences the physical characteristics of meat. Dielectric permittivity and loss factor values as a function of temperature for beef, turkey, ham, and so forth are represented in Tables 3.4 to 3.9.

Table 3.4
Dielectric Properties of Beef (Shoulder Part)

Temperature	Frequency (MHz)	Expression	R^2
$30 \leq T°C \leq 90$	433	$\varepsilon'(T) = 4 \cdot 10^{-5}T^3 - 0.006T^2 + 0.1397T + 52.726$	0.9952
		$\varepsilon''(T) = -0.0045T^2 + 0.4468T + 27.012$	0.9964
	915	$\varepsilon'(T) = -4 \cdot 10^{-5}T^3 + 0.0085T^2 - 0.7389T + 67.025$	0.9980
		$\varepsilon''(T) = -0.0043T^2 + 0.4504T + 20.221$	0.9810
	2450	$\varepsilon'(T) = -10^{-6}T^3 + 0.0003T^2 - 0.1717T + 51.269$	0.9974
		$\varepsilon''(T) = -0.0037T^2 + 0.3266T + 14.182$	0.9959

From: [3].

Table 3.5
Comparison of Dielectric Properties of Beef at 2,450 MHz
$5 \leq T°C \leq 95$

Beef	Expression	R^2
Fresh, W = 75.4%	$\varepsilon'(T) = -4 \cdot 10^{-5}T^3 + 0.0052T^2 - 0.3471T + 51.561$	0.9972
	$\varepsilon''(T) = 8 \cdot 10^{-6}T^3 - 0.0031T^2 + 0.2483T + 13.438$	0.9731
Cooked, W = 72.4%	$\varepsilon'(T) = -10^{-5}T^3 + 0.0017T^2 - 0.1715T + 49.806$	0.9902
	$\varepsilon''(T) = 2 \cdot 10^{-5}T^3 - 0.0031T^2 + 0.1895T + 7.0384$	0.9851

From: [3].

Table 3.6
Dielectric Properties of Some Meat Samples in Temperature Range
$40 \leq T°C \leq 100$ at 433 MHz

Sample	W %	Fat content %	Expression	R^2
Ham	69.1	4.7	$\varepsilon'(T) = 30.293\ln(T) - 51.735$	0.9970
			$\varepsilon''(T) = 15.535\ln(T) - 13.498$	0.9919
Liver paste	55	20	$\varepsilon'(T) = -0.0008T^2 + 0.0628T + 39.77$	0.9935
			$\varepsilon''(T) = -0.0007T^2 + 0.149T + 10.788$	0.9838

From: [3].

Table 3.7
Dielectric Properties of Raw Meat in Temperature Range $5 \leq T°C \leq 65$

Raw Meat	f (MHz)	Expression	R^2
Beef	915	$\varepsilon'(T) = 57.39 - 0.10716T$	0.9999
		$\varepsilon''(T) = 17.333exp(0.0099T)$	0.9987
	2450	$\varepsilon'(T) = 53.2 - 0.7133T$	0.9999
		$\varepsilon''(T) = 3 \cdot 10^{-5}T^3 - 0.0005T^2 - 0.1087T + 17.762$	0.9367
Turkey	915	$\varepsilon'(T) = 60.358 - 0.071666T$	0.9999
		$\varepsilon''(T) = 22.217 + 0.43666T$	0.9999
	2450	$\varepsilon'(T) = 2 \cdot 10^{-5}T^3 - 0.0032T^2 + 0.0854T + 52.499$	0.9944
		$\varepsilon''(T) = -9 \cdot 10^{-5}T^3 + 0.013T^2 - 0.4955T + 27.622$	0.9936

From: [6].

Table 3.8
Dielectric Properties of Meat at 2,450 MHz

Meat	Temperature	Expression	R^2
Chicken breast	$15 \leq T°C \leq 65$	$\varepsilon'(T) = 2 \cdot 10^{-5}T^3 - 0.0034T^2 + 0.1126T + 54.969$	0.9957
		$\varepsilon''(T) = 0.0018T^2 - 0.0976T + 19.479$	0.9982
Chicken thigh	$15 \leq T°C \leq 55$	$\varepsilon'(T) = -0.0013T^2 + 0.012T + 56.039$	0.9982
		$\varepsilon''(T) = 0.0019T^2 - 0.1119T + 17.301$	0.9964
Beef	$15 \leq T°C \leq 60$	$\varepsilon'(T) = 3 \cdot 10^{-5}T^3 - 0.0051T^2 + 0.1437T + 53.616$	0.9983
		$\varepsilon''(T) = 0.0018T^2 - 0.1216T + 18.363$	0.9831

From: [7].

Dielectric properties of meat were measured using the waveguide method [3] and the open-ended coaxial probe method [7, 8]. In [8], one can find predictive equations for the CDP as a function of temperature, moisture, and ash at 2,450 MHz for various fruits, vegetables, and meat samples (Table 3.9).

Table 3.9
Dielectric Properties of Ash Turkey at 2,450 MHz
in Temperature Range $25 \leq T°C \leq 125$

Ash%	Expression	R^2
6	$\varepsilon'(T) = 10^{-5}T^3 - 0.0032T^2 + 0.1142T + 54.516$	0.9971
	$\varepsilon''(T) = -10^{-5}T^3 + 0.0023T^2 + 0.0907T + 20.428$	0.9966
15	$\varepsilon'(T) = -8 \cdot 10^{-6}T^3 + 0.0022T^2 - 0.177T + 52.757$	0.6617
	$\varepsilon''(T) = 0.0032T^2 + 0.16T + 33$	0.9849

From: [8].

According to [9], the CDP of ground beef is determined as:

$$\varepsilon' - j\varepsilon'' = u_0 + u_1 T + u_2 T^2 + u_3 (FT) + u_4 F + u_5 F^2 + u_6 T^{-1} \qquad (3.1)$$

where F is the fat content and the values of coefficients $u_0 \div u_6$ are given in Table 3.10.

Table 3.10
Dielectric Properties of Ground Beef with Fat Content $4 < F\% < 20$ at 915 MHz and 2.45 GHz in Temperature Range $-20 < T°C < 74$

DP	f (MHz)	$T°C$	u_0	u_1	u_2	u_3	u_4	u_5	u_6	R^2
ε'	915	$< T_f$	−3.07	−0.24	NS	0.026	1.24	−0.03	−12.73	0.87
		$> T_f$	57.08	−0.1	0.0009	−0.001	−1.49	0.045	NS	0.94
	2,450	$< T_f$	−4.19	−0.24	NS	0.021	1.33	−0.035	−12.62	0.86
		$> T_f$	50.69	−0.08	0.0005	NS	−1.23	0.03	NS	0.95
ε''	915	$< T_f$	3.12	0.66	0.2	0.009	0.59	−0.2	NS	0.86
		$> T_f$	20.34	0.05	0.001	−0.002	−0.88	0.03	NS	0.93
	2,450	$< T_f$	−2.17	NS	NS	0.004	0.57	−0.02	−4.93	0.87
		$> T_f$	16.7	−0.05	0.0007	NS	0.22	NS	NS	0.83

From: [9].
NS = not significant; T_f = freezing point.

Thermal properties of meat are also well described in the literature [10, 12]. Along with meat, animal fats are widely used in the food industry. Temperature dependencies of the main thermal parameters of animal fats are shown in Table 3.11. According to [6], the CDP of bacon fat, for example, at room temperature is quite low: $\varepsilon' - j\varepsilon'' = 2.6 - j0.16$ (1 GHz); $\varepsilon' - j\varepsilon'' = 2.5 - j0.13$ (3 GHz). Information about $\rho_t(T)$, $C_t(T)$, $\lambda_t(T)$ in Table 3.11 has been collected in [10] from different sources both at low and high temperatures. Deviation of some parameters is explained by different measuring methods and volume fractions of free and bound water, salt, and other chemical substances content in food samples. The same is true for the raw meat parameters represented in Table 3.12.

Table 3.11
Thermal Properties of Animal Fats

Fat	Temperature	Parameters	Units
Raw beef fat	$278 \leq T°K \leq 303$	$\rho_t(T) = 1377 - 1.5T$	kg/m³
	$313 \leq T°K \leq 363$	$\rho_t(T) = 1076 - 0.5T$	
	$75 \leq T°K \leq 273$	$C_t(T) = 308.37exp(0.0076T); R^2 = 0.9805$	J/(kg·K)
	$298 \leq T°K \leq 338$	$C_t(T) = 0.3048T^3 - 292.79T^2 + 93538T - 10^7$ $R^2 = 0.9802$	
	$243 \leq T°K \leq 273$	$\lambda_t(T) = 0.8888 - 0.0025T, R^2 = 0.9952$	W/(m·K)
	$313 \leq T°K \leq 363$	$\lambda_t(T) = -9 \cdot 10^8 T^3 + 10^{-4}T^2 - 0.0341T + 4.175$ $R^2 = 0.9905$	
Raw mutton fat	$323 \leq T°K \leq 363$	$\rho_t(T) = 1124 - 0.75T$	kg/m³
	$283 \leq T°K \leq 343$	$C_t(T) = -0.0028T^3 - 0.3329T^2 + 998.74T - 187932; R^2 = 0.8957$	J/(kg·K)
	$254 \leq T°K \leq 311$	$\lambda_t(T) = 10^{-6}T^3 - 0.0009T^2 + 0.2544T - 22.969$ $R^2 = 0.9957$	W/(m·K)
Salted pork fat	$77 \leq T°K \leq 273$	$\rho_t(T) = 1052 - 0.377T$	kg/m³
	$283 \leq T°K \leq 333$	$\rho_t(T) = 1133 - 0.63T$	
	$333 \leq T°K \leq 353$	$\rho_t(T) = 1642 - 2.21T$	
	$283 \leq T°K \leq 298$	$C_t(T) = 5.64T^2 - 3144.4T + 440897$	J/(kg·K)
	$303 \leq T°K \leq 338$	$C_t(T) = -0.1235T^3 + 121.03T^2 - 39509T + 4 \cdot 10^6; R^2 = 0.971$	
	$243 \leq T°K \leq 273$	$\lambda_t(T) = -10^{-4}T^2 + 0.0498T - 5.6071$ $R^2 = 0.9968$	W/(m·K)
Grease	$323 \leq T°K \leq 373$	$\rho_t(T) = 1069 - 0.555T$	kg/m³
	$75 \leq T°K \leq 253$	$C_t(T) = 0.0005T^3 - 0.2027T^2 + 30.72T - 691.92; R^2 = 0.9864$	J/(kg·K)
	$253 \leq T°K \leq 273$	$2850 \leq C_t(T) \leq 5240$	
	$283 \leq T°K \leq 298$	$C_t(T) = 7.96T^2 - 4484.1T + 634518$ $R^2 = 0.9995$	
	$303 \leq T°K \leq 333$	$C_t(T) = -0.3524T^3 + 341.48T^2 - 110253T + 10^7;$ $R^2 = 0.9777$	
	$77 \leq T°K \leq 323$	$\lambda_t(T) = 0.65 - 0.0015T$	W/(m·K)
	$323 \leq T°K \leq 373$	$\lambda_t(T) = 0.031 + 0.0004T$	

From: [10].

Sometimes the mathematical function of temperature dependence of this or that thermal parameter is not known but one can mention minimum and maximum values of a given parameter in some temperature range. Several examples of such an approach are represented in Table 3.12.

Table 3.12
Thermal Properties of Raw Meat

Parameter	Units	$T°K$	Value	Comment
Thermal conductivity	W/(m·K)	293 ÷ 333	$\lambda_t(T) = 0.422 + 0.00019T$	Beef
			$\lambda_t(T) = 0.430 + 0.00016T$	Mutton
			$\lambda_t(T) = 0.250 + 0.00087T$	Pork
Thermal diffusivity ×10⁸	m²/s	273 ÷ 303	11.7 ÷ 12.5	Beef (W = 74%) and pork (W = 76.8%)
		293 ÷ 343	12.2 ÷ 12.5	Beef
		380 ÷ 430	$a_t(T) = 6 \cdot 10^{-5}T^3 - 0.0822T^2 + 37.509T - 5595.7$	$R^2 = 0.9806$
Density	kg/m³	278 ÷ 303	1066 ÷ 1130	Beef
			1070	Pork
			1265 ÷ 1247	Bird
Heat capacity	J/(kg·K)	273 ÷ 285	3700 ÷ 4100	Beef
		285 ÷ 310	4100 ÷ 3590	
		310 ÷ 320	3590 ÷ 3570	
		320 ÷ 373	3056 ÷ 3224	

From: [10].

3.2   Fish

Microwave energy is widely utilized for defrosting, tempering, heating, and sterilization of fish and marine products. Most publications on dielectric properties of fish contain information about $\varepsilon' - j\varepsilon''$ at room temperature (Tables 3.13 and 3.14). Both dielectric permittivity and loss factor mainly depend on moisture and fat content at a fixed temperature; moreover, sometimes density should also be taken into account (Table 3.14). Temperature dependencies of CDP for different fish and caviar samples are shown in Tables 3.15 to 3.17.

Table 3.13
Dielectric Properties of Some Fish Samples
at 2,450 MHz and $T = 20°C$

Fish	Water Content (%)	Fat content (%)	ε'	ε''
Herring	61.3	17.4	42	10.2
Cod	76.3	0.5	52	18
Paltus (halibut)	75	5	48	17.5
Hake	74.5	1.5	43.3	17.8
Tuna	72.3	3.4	42	15
Soodak (pike-perch)	69.9	1.9	44	16

From: [3].

Table 3.14
Dielectric Properties of Hake Stuffing at $T = 20°C$

Density (kg/m³)	433 MHz		915 MHz		2450 MHz	
	ε'	ε''	ε'	ε''	ε'	ε''
920	49.3	31.1	45.7	25.5	40.5	16.8
960	51.1	31.5	49.9	26.1	43	17.4
990	54	32.4	52.3	26.6	47.3	18.1
1,010	56.8	33.2	54	27.5	49.6	18.7
1,025	57.6	33.7	56.7	28.3	51	19.3

From: [3].

Table 3.15
Dielectric Properties of Cod and Paltus (Halibut) at 433 MHz in
Temperature Range $10 \leq T°C \leq 100$

Fish	Expression	R^2
Cod	$\varepsilon'(T) = -0.0025129T^2 + 0.1154T + 56.255$	0.9993
	$\varepsilon''(T) = -2 \cdot 10^{-5}T^3 - 2 \cdot 10^{-5}T^2 + 0.258T + 29.186$	0.9979
Paltus (halibut)	$\varepsilon'(T) = -0.0015T^2 - 0.0192T + 54.997$	0.9993
	$\varepsilon''(T) = -10^{-5}T^3 - 0.001T^2 + 0.2741T + 25.408$	0.9991

From: [3].

Complex dielectric permittivity of cooked (boiled) cod at 2,450 MHz in temperature range $20 \leq T°C \leq 100$ [3]:

$$\varepsilon'(T) = 48.0825 - 0.07125T, \; R^2 = 0.9999 \tag{3.2}$$

$$\varepsilon''(T) = -0.00117T^2 + 0.2012T + 4.5684, \; R^2 = 0.9988 \tag{3.3}$$

Table 3.16
Dielectric Properties of Fish at 2,450 MHz

Fish	Temperature	Expression	R^2
Cod	$15 \leq T°C \leq 50$	$\varepsilon'(T) = -0.0011T^2 - 0.0493T + 63.548$	0.9985
		$\varepsilon''(T) = -7 \cdot 10^{-5}T^3 + 0.0093T^2 - 0.3828T + 21.289$	0.9945
Salmon	$15 \leq T°C \leq 50$	$\varepsilon'(T) = 53.583\exp(-0.0018T)$	0.9193
		$\varepsilon''(T) = -8 \cdot 10^{-5}T^3 + 0.009T^2 - 0.2997T + 19.249$	0.9547
Perch	$15 \leq T°C \leq 40$	$\varepsilon'(T) = 61.616\exp(-0.0021T)$	0.9854
		$\varepsilon''(T) = 0.0028T^2 - 0.1491T + 20.193$	0.9759

From: [7].

Table 3.17
Dielectric Properties of Fish Products at 915 MHz in Temperature Range
$20 \leq T°C \leq 80$

Fish product	Expression	R^2
Unsalted caviar	$\varepsilon'(T) = 0.0004T^2 + 0.0035T + 32.25$	0.9879
	$\varepsilon''(T) = -6 \cdot 10^{-5}T^3 + 0.0099T^2 - 0.3867T + 13.15$	0.9947
Salted caviar	$\varepsilon'(T) = -5 \cdot 10^{-5}T^3 + 0.0097T^2 - 0.4387T + 30.379$	0.9968
	$\varepsilon''(T) = 24.64\exp(0.0177T)$	0.9965
Unsalted salmon	$\varepsilon'(T) = 0.0071T^2 - 0.8767T + 44.414$	0.9627
	$\varepsilon''(T) = -4 \cdot 10^{-5}T^3 + 0.0131T^2 - 0.9021T + 31.736$	0.9894
Salted salmon	$\varepsilon'(T) = 0.0054T^2 - 0.5945T + 38.864$	0.8869
	$\varepsilon''(T) = -2 \cdot 10^{-4}T^3 + 0.0405T^2 - 2.1038T + 68.114$	0.9954

From: [11].

Data about thermal characteristics of various marine products taken from [10] and [12] are shown in Tables 3.18 to 3.23. Parameters X_f and X_w in Table 3.19 are the fat and moisture content (dry basis), respectively. As noted in [10], the thermal parameters of fish stuffing are approximately the same as for the fish itself. Moisture, fat content, and temperature are again general factors defining thermal conductivity, heat capacity, and density of fish.

Table 3.18
Thermal Conductivity of Fish

Fish	Temperature	Moisture	Thermal Conductivity (W/(m·K))
Frozen tuna	243°K	$70 \leq W\% \leq 73$	$\lambda_t = 1.417$
	253°K		$\lambda_t = 1.333$
	263°K		$\lambda_t = 1.167$
	273°K		$\lambda_t = 0.444$
Bream	$273 \leq T°K \leq 303$	$75 \leq W\% \leq 78$	$\lambda_t = 0.47$
Mallotus villosus	$278 \leq T°K \leq 303$	$78 \leq W\% \leq 80$	$\lambda_t(T) = 0.003T - 0.449$
Cod	$275 \leq T°K \leq 293$	$75 \leq W\% \leq 79$	$0.45 \leq \lambda_t \leq 0.54$
	$293 \leq T°K \leq 313$		$\lambda_t(T) = 0.001T + 0.18$
	$323 \leq T°K \leq 373$		$\lambda_t(T) = 0.002T - 0.15$
Pike-perch	$273 \leq T°K \leq 288$	$76 \leq W\% \leq 80$	$0.47 \leq \lambda_t \leq 0.52$
Tuna	$273 \leq T°K \leq 278$	$73 \leq W\% \leq 76$	$0.5 \leq \lambda_t \leq 0.58$
Salmon	$293 \leq T°K \leq 299$	$75 \leq W\% \leq 79$	$0.54 \leq \lambda_t \leq 0.56$
Mackrel	$318 \leq T°K \leq 323$		$0.57 \leq \lambda_t \leq 0.6$
	$283 \leq T°K \leq 373$	$78 \leq W\% \leq 80$	$\lambda_t(T) = -5 \cdot 10^{-7}T^3 + 0.0005T^2 - 0.1464T + 15,783; R^2 = 0.9915$

From: [10].

Table 3.19
Thermal Conductivity λ_t, [W/(m·K)], and Diffusivity $a_t \times 10^7$ [m^2/s] of Fish
at $0 \leq T°C \leq 40$

Fish	X_f	X_w	Expression	R^2
Salmon	0.0046	0.77	$\lambda_t(T) = 2 \cdot 10^{-6}T^3 - 8 \cdot 10^{-5}T^2 + 0.0016T + 0.4091$	0.9997
			$a_t(T) = -5 \cdot 10^{-5}T^3 + 0.0031T^2 - 0.0284T + 1.6746$	0.9981
Sole	0.0046	0.77	$\lambda_t(T) = 2 \cdot 10^{-6}T^3 - 0.0002T^2 + 0.0067T + 0.3956$	0.8844
			$a_t(T) = 2 \cdot 10^{-5}T^3 - 0.0009T^2 + 0.0328T + 1.9026$	0.9999
Hilsa	0.05	0.463	$\lambda_t(T) = -10^{-6}T^3 + 8 \cdot 10^{-5}T^2 - 10^{-4}T + 0.4092$	0.9991
			$a_t(T) = -5 \cdot 10^{-5}T^3 + 0.0026T^2 - 0.0097T + 2.1285$	0.8975

From: [12].

Table 3.20
Thermal Conductivity of Fish Mince, Sausages, and Caviar

Product	ρ_t (kg/m³)	Temperature	λ_t (W/(m·K))
Pollock mince	940	283°K	$0.63 \leq \lambda_t \leq 0.7$
	980	$273 \leq T°K \leq 343$	$0.45 \leq \lambda_t \leq 0.47$
		$343 \leq T°K \leq 373$	$0.75 \leq \lambda_t \leq 1.3$
Pike mince	995	$353 \leq T°K \leq 401$	$0.46 \leq \lambda_t \leq 0.57$
Perch mince			$0.41 \leq \lambda_t \leq 0.43$
Cod mince			$0.43 \leq \lambda_t \leq 0.52$
Pike sausage	970	$353 \leq T°K \leq 401$	$0.43 \leq \lambda_t \leq 0.49$
Perch sausage			$0.38 \leq \lambda_t \leq 0.45$
Black caviar	900	$269 \leq T°K \leq 333$	$\lambda_t(T) = 0.442 - 0.0139(T{-}273) + 1.3 \cdot 10^{-5}(T{-}273)^2$

From: [10].

Table 3.21
Heat Capacity of Fish

Fish	Temperature	Moisture	Heat Capacity (J/(kg·K))
Cod filet	$20 \leq T°C \leq 25$	$78 \leq W\% \leq 80.3$	$3,434 \leq C_t \leq 3,740$
	$2 \leq T°C \leq 18$	80.3%	$3,642 \leq C_t \leq 3,684$
Sheatfish	$4 \leq T°C \leq 25$	78%	$3,433 \leq C_t \leq 3,685$
Sazan	$20 \leq T°C \leq 25$	$74 \leq W\% \leq 78$	$3,659 \leq C_t \leq 3,726$
Pike-perch	$0 \leq T°C \leq 25$	$76.8 \leq W\% \leq 78.4$	$3,538 \leq C_t \leq 3,798$
Sea perch	$20 \leq T°C \leq 25$	79.1%	3,600
Sturgeon	$20 \leq T°C \leq 25$	70.8%	3,643
Haddock	$2 \leq T°C \leq 18$	83.6%	$3,684 \leq C_t \leq 3,726$

From: [10].

Table 3.22
Density of Fish at 15°C

Fish	Mass (g)	Density (kg/m³)
Herring	$390 \leq m \leq 726$	$1066 \leq \rho_t \leq 1109$
Bream	$386 \leq m \leq 811$	$1015 \leq \rho_t \leq 1033$
Pike-perch	$350 \leq m \leq 3400$	$955 \leq \rho_t \leq 996$
River perch	$228 \leq m \leq 369$	$987 \leq \rho_t \leq 1097$

From: [10].

Table 3.23
Thermal Parameters of Fish

Fish	$T°$K	W%	ρ_t (kg/m³)	C_t (J/(kg·K))	λ_t (W/(m·K))
Cod	293	80	1,020	3,684	0.46
		83	997	3,684	0.54
Salmon	293	73	980	3,517	0.5
Pike-perch	273	80	1,070	3,475	0.52
	293	80	1,064	3,810	0.47
Sazan	293	80	1,060	3,864	0.44
Sturgeon	293	80	1,059	3,643	0.43

From: [10].

3.3   Eggs

Dielectric and thermal parameters of eggs are represented in Tables 3.24 to 3.26.

Table 3.24
Dielectric and Thermal Properties of Eggs at 915 MHz in
Temperature Range $20 \leq T°C \leq 60$

Product	Parameter	Expression	Units
Egg white	ε'	$-0.257T + 75.6$	
	ε''	$0.117T + 19.6$	
	λ_t	$0.00096T + 0.545$	W/(m·K)
	C_t	$0.761T + 3890$	J/(kg·K)
	ρ_t	$-0.567T + 1047$	kg/m³
Egg yolk	ε'	$-0.149T + 35.5$	
	ε''	$-0.006T + 11.1$	
	λ_t	$0.00056T + 0.38$	W/(m·K)
	C_t	$1.75T + 2930$	J/(kg·K)
	ρ_t	$-0.641T + 1042$	kg/m³

Personal communication. Wang, S., WSU, WA, May 2003.

Table 3.25
Dielectric Properties of Eggs at 2,450 MHz and Different Temperatures

	20		40		60		80	
$T°C$	ε'	ε''	ε'	ε''	ε'	ε''	ε'	ε''
Egg white	46	4.5	40.8	3.75	37.5	2.5	35	2.5
Egg yolk	52.5	8.75	47.5	7.5	44.3	6.25	40.5	6.3

From: [3].

Table 3.26
Thermal Properties of Eggs

Egg part	Temperature	Parameters	Units
White	$0 \leq T°C \leq 80$	$\rho_t(T) = -2 \cdot 10^{-5}T^3 + 0.0001T^2 - 0.2699T$ $+ 1047.8; R^2 = 0.998$	kg/m^3
Yolk		$\rho_t(T) = -0.0001T^3 + 0.015T^2 - 0.9476T +$ $+ 1039.1; R^2 = 0.9766$	
Mélange		$\rho_t(T) = 3 \cdot 10^{-5}T^3 - 0.0076T^2 - 0.0384T + 1041.7$ $R^2 = 0.9928$	
White W=80%	$2.5 \leq T°C \leq 37.5$	$C_t(T) = 0.4524T^2 - 16.143T + 3725$	J/(kg·K)
	$278 \leq T°K \leq 323$	$\lambda_t(T) = 0.117 + 0.00142T$	W/(m·K)
Yolk W=50%	$273 \leq T°K \leq 323$	$C_t(T) = 2326 + 2.2T$	J/(kg·K)
	$279 \leq T°K \leq 323$	$\lambda_t(T) = 0.3455 + 0.0002T$	W/(m·K)
Mélange	$273 \leq T°K \leq 323$	$3433 \leq C_t(T) \leq 3688$	J/(kg·K)
		$\lambda_t(T) = 0.11 + 0.00125T$	W/(m·K)

From: [10].

3.4   Cheese, Butter, and Milk

Physical properties of milk, cheese, and butter depend on several factors as shown in Tables 3.27 to 3.33.

Table 3.27
Dielectric Properties of Milk at 2,450 MHz and $T = 20°C$

Milk product	Fat (%)	Protein (%)	Lactose (%)	Moisture (%)	ε'	ε''
1% Milk [13]	0.94	3.31	4.93	90.11	70.56	17.61
3.25%Milk [13]	3.17	3.25	4.79	88.13	67.98	17.63
Milk [3]	0	—	—	81	69.8	15.7
Milk [3]	0	—	—	87	66.4	17.3
Milk [3]	0	—	—	83	63	18.8
Milk fat (solid)	100	0	0	0	2.613	0.153

Table 3.28
Dielectric Properties of Cottage Cheese at 915 MHz and 2.45 GHz

f (MHz)	$T°C$	Fat-Free		Fat Content = 2%		Fat Content = 4%	
		ε'	ε''	ε'	ε''	ε'	ε''
915	5	67.4	32.9	62.6	26.8	57.7	26.2
	25	63.5	36.7	62.8	31.9	55.9	30.1
	45	64.5	43.5	61.9	38.6	57.2	38.6
	65	55.2	51.2	54.9	58.2	50.4	55.1
2,450	5	61.3	26	58.1	22.7	52.7	21
	25	59.9	22.4	59.8	20.6	54.6	19.5
	45	62.5	24.2	60	21.2	55.8	20.9
	65	52.4	23.1	52.6	24.8	50.2	24.2

From: [14].

Table 3.29
Dielectric Properties of Butter and Albumen

Foodstuff	Temperature	f (GHz)	Dielectric Permittivity and Loss Factor
Albumen–milk concentrate [15]	$20 < T°C < 70$	2.45	$\varepsilon'(T) = 55 - 0.17T$
			$\varepsilon''(T) = 17.89 - 0.1T$
Butter [3]	$5 < T°C < 40$	2.45	$\varepsilon'(T) = 0.0002T^2 + 0.0052T + 4.1567$ $R^2 = 0.9962$
			$\varepsilon''(T) = -9 \cdot 10^{-6}T^3 + 0.0007T^2 - 0.0099T + 0.556$ $R^2 = 0.982$
Butter [6]	$5 < T°C < 40$	2.45	$\varepsilon'(T) = 2 \cdot 10^{-5}T^3 - 0.0011T^2 + 0.0179T + 3.9886$ $R^2 = 0.9932$
			$\varepsilon''(T) = 5 \cdot 10^{-6}T^3 - 0.0002T^2 - 0.0038T + 0.3822$ $R^2 = 0.9965$

Whey protein mixture/gel (see Table 3.30 for the dielectric properties) is a specific substance used in food science to study the uniformity of microwave heating by means of a chemical marker method [16]. It consists of whey powder, sodium chloride, glucose, and distilled water.

Microwaves are often employed for thermal treatment of mixed products, like macaroni and cheese or cheese sauce prepared by mixing melted margarine, cheese powder with milk, and water [16].

The CDP of all these foods is represented in Table 3.30.

Table 3.30
Dielectric Properties of Some Foods at 915 and 1,800 MHz
in Temperature Range $20 \leq T°C \leq 121$

Food	f (MHz)	Expression	R^2
Whey protein gel	915	$\varepsilon'(T) = 0.0004T^2 - 0.1494T + 61.986$	0.9945
		$\varepsilon''(T) = 0.0015T^2 + 0.3963T + 26.008$	0.9998
	1,800	$\varepsilon'(T) = 59.386 - 0.121T$	0.9969
		$\varepsilon''(T) = 19.417exp(0.0081T)$	0.9988
Liquid whey protein mixture	915	$\varepsilon'(T) = 5 \cdot 10^{-6}T^3 - 0.0004T^2 - 0.1445T + 64.854$	0.9973
		$\varepsilon''(T) = 0.0017T^2 + 0.4172T + 24.231$	0.9997
	1,800	$\varepsilon'(T) = 0.0004T^2 - 0.184T + 62.833$	0.9971
		$\varepsilon''(T) = 19.12exp(0.0087T)$	0.9991
Cheese sauce	915	$\varepsilon'(T) = 0.0002T^2 - 0.1298T + 45.096$	0.9966
		$\varepsilon''(T) = -9 \cdot 10^{-6}T^3 + 0.0032T^2 + 0.6006T + 32.466$	0.9996
	1,800	$\varepsilon'(T) = 10^{-5}T^3 - 0.0024T^2 + 0.0614T + 38.949$	0.9966
		$\varepsilon''(T) = 0.0009T^2 + 0.3036T + 20.916$	0.9993
Macaroni and cheese	915	$\varepsilon'(T) = 40.871 - 0.0081T$	0.7381
		$\varepsilon''(T) = 6 \cdot 10^{-6}T^3 - 4 \cdot 10^{-6}T^2 + 0.2637T + 15.898$	0.9994
	1,800	$\varepsilon'(T) = 40.201 - 0.0328T$	0.8826
		$\varepsilon''(T) = 0.0009T^2 + 0.0119T + 16.814$	0.9991

From: [16].

Thermal characteristics of milk as a function of temperature and fat content (n) are listed in Tables 3.31 and 3.32. Less information is available in the literature about temperature-dependent thermal properties of cheese and butter. One example of the heat capacity of cheese is shown in Table 3.31. Whey protein gel properties were defined in [17] up to the temperature $T = 80°C$ (Table 3.33) necessary for providing pasteurization processing of food [2].

Table 3.31
Heat Capacity of Cheese and Milk

Food	Temperature	Fat Content (Dry Basis)	Heat Capacity, J/(kg·K)
Cheese [12]	$40 \leq T°C \leq 100$	0.135	$C_t(T) = 1.65T + 2679.8$
		0.345	$C_t(T) = 1.3179T + 3026.5$
Milk [10]	$273 \leq T°K \leq 308$	Fat-free	$C_t(T) = 15.9T - 3353$
	$313 \leq T°K \leq 353$	Fat-free	$C_t(T) = 11.3T - 1712$
	$303 \leq T°K \leq 353$	Fat-free	$C_t(T) = 16.8T - 3242$

Table 3.32
Thermal Properties of Milk

Parameter	Units	$T°K$	Value	Comment
Thermal conductivity	W/(m·K)	293	$0.52 \div 0.59$	Different fat content
		353	$0.612 \div 0.643$	
Density	kg/m^3	$274 \div 343$	$\rho_t(T) = 1055 - 0.179T + 3.14n$	$0.082 \leq n \leq 0.102$
Thermal diffusivity $\times 10^8$	m^2/s	273	$12.6 \div 12.8$	Different fat content
		293	$13.4 \div 13.5$	
		313	$14 \div 14.2$	
		333	$14.6 \div 14.8$	
		353	$15.1 \div 15.3$	

From: [10].

Table 3.33
Thermal Properties of Whey Protein Gel

Parameter	Temperature	Expression	R^2
Volumetric specific heat MJ/(m^3·K)	$0 \leq T°C < 50$	$C_t(T) \cdot \rho_t(T) = -0.0002T^2 + 0.0095T + 3.8444$	0.926
	$50 \leq T°C < 80$	$C_t(T) \cdot \rho_t(T) = -0.0004T^2 + 0.0521T + 2.2407$	0.999
Thermal conductivity W/(m·K)	$0 \leq T°K \leq 80$	$\lambda_t(T) = 0.0009T + 0.5158$	0.993

From: [17].

3.5   Vegetables, Fruits, and Nuts

Microwaves are successfully used for postharvest treatment, drying, pasteurization, sterilization, defrosting, and cooking of such agricultural products as vegetables and fruits [2, 3]. Chemical substances, moisture, and temperature influence much on CDP of many agricultural foods. Moreover physical properties of fresh and cooked vegetables and fruits are different. Potato is one of the most popular foods in the world and information about its properties is available in the literature. Temperature dependences of CDP of both fresh and cooked potato are given in Tables 3.34 to 3.36. Dielectric properties of different vegetables and fruits are represented in Tables 3.37 and 3.38. Predictive equations in Tables 3.35 and 3.38 have been obtained in [18] and [19], respectively. The open-ended coaxial probe method was used in [18–20].

Table 3.39 shows dielectric permittivity and loss factor of some fresh fruits and nuts measured in [20] at 915 and 1,800 MHz. Additional data on given parameters at 27 and 40 MHz can also be found in [20].

Table 3.34
Dielectric Properties of Potato at $20 \leq T°C \leq 95$ and $W = 77.8\%$

Frequency (MHz)	Expression	R^2
433	$\varepsilon'(T) = -10^{-6}T^3 - 0.0003T^2 - 0.1387T + 69.332$	0.9968
	$\varepsilon''(T) = 2 \cdot 10^{-5}T^3 - 0.0079T^2 + 0.6603T + 22.692$	0.9951
915	$\varepsilon'(T) = -0.0008T^2 - 0.1118T + 68.005$	0.9985
	$\varepsilon''(T) = -8 \cdot 10^{-7}T^3 - 0.0024T^2 + 0.3024T + 21.483$	0.9893
2450	$\varepsilon'(T) = -0.0011T^2 - 0.0834T + 65.3$	0.9994
	$\varepsilon''(T) = 10^{-5}T^3 - 0.0035T^2 + 0.3012T + 12.085$	0.9468

From: [3].

Table 3.35
Dielectric Properties of Mashed Potatoes at 433 and 915 MHz in
Temperature Range $20 \leq T°C \leq 121$

f (MHz)	Expression	R^2
433	$\varepsilon' = -68.4 + 0.94M + 115S - 0.00138T \cdot M - 61.6S^2 + 9.42S^3$	0.92
	$\varepsilon'' = 14.9 + 0.583T \cdot S$	0.91
915	$\varepsilon' = -90.2 + 1.17M + 112S - 0.087T \cdot S - 10^{-5}T^2 \cdot M + 6.72 \cdot 10^{-4}T^2S$ $+ 59.6S^2 + 9.16S^3$	0.91
	$\varepsilon'' = 12.4 + 0.267T \cdot S$	0.88

From: [18].
M is the moisture content (%,w.b.); S is the salt content (%,w.b.).

Table 3.36
Dielectric Properties of Cooked (Boiled)
Potato with W = 78.7% at 433 MHz and
W = 79.2% at 915 MHz

Temperature (°C)	433 MHz		915 MHz	
	ε'	ε''	ε'	ε''
20	54.2	57.6	53.3	20.8
60	49.3	70.4	49.2	25.0
80	44.1	82.9	45.8	28.8
100	42.4	99.5	42.6	34.8

From: [3].

Equations in Table 3.38 describe only temperature dependences of vegetable and fruit properties. But authors [19] also propose some kind of generalized expressions for CDP of all studied agricultural products as a function of temperature, moisture, and ash content.

Table 3.37
Dielectric Properties of Some Vegetables at 2,450 MHz
in Temperature Range $20 \leq T°C \leq 100$

Vegetable	Expression	R^2
Carrot W = 88.2%	$\varepsilon'(T) = 0.001T^2 - 0.3166T + 79.007$	0.9991
	$\varepsilon''(T) = 0.0012T^2 - 0.1993T + 17.018$	0.9843
Pease W = 82.6%	$\varepsilon'(T) = -0.2064T + 66.398$	0.9967
	$\varepsilon''(T) = 0.0008T^2 - 0.1332T + 13.739$	0.8841

From: [3].

Table 3.38
Dielectric Properties of Vegetables and Fruits at 2.45 GHz
in Temperature Range $5 \leq T°C \leq 120$

Vegetable and Fruit	Expression	R^2
Broccoli	$\varepsilon'(T) = -0.0005129T^2 - 0.1208T + 74.18$	0.989
	$\varepsilon''(T) = 0.001174T^2 - 0.0242T + 20.89$	0.915
Carrot	$\varepsilon'(T) = -0.2068T + 77.94$	0.906
	$\varepsilon''(T) = 0.0016T^2 - 0.104T + 21.68$	0.927
Garlic	$\varepsilon'(T) = -0.001143T^2 + 0.1743T + 43.47$	0.848
	$\varepsilon''(T) = 0.00125T^2 - 0.09577T + 19.62$	0.957
Parsnip	$\varepsilon'(T) = -0.1323T + 65.31$	0.928
	$\varepsilon''(T) = 0.001332T^2 - 0.1385T + 22.4$	0.893
Radish	$\varepsilon'(T) = -0.2166T + 77.31$	0.909
	$\varepsilon''(T) = 0.001517T^2 - 0.1825T + 20.36$	0.846
Turnip	$\varepsilon'(T) = -0.1528T + 66.93$	0.727
	$\varepsilon''(T) = 0.001471T^2 - 0.1965T + 17.99$	0.907
Apple	$\varepsilon'(T) = -0.0005831T^2 - 0.0524T + 64.63$	0.960
	$\varepsilon''(T) = 0.001348T^2 - 0.2407T + 17.23$	0.974
Banana	$\varepsilon'(T) = -0.0005706T^2 - 0.08229T + 68.73$	0.994
	$\varepsilon''(T) = 0.00145T^2 - 0.1536T + 21.28$	0.895
Corn	$\varepsilon'(T) = -0.0005502T^2 - 0.03792T + 60.86$	0.950
	$\varepsilon''(T) = 0.001004T^2 - 0.1355T + 19.02$	0.883
Cucumber	$\varepsilon'(T) = -0.2136T + 77.82$	0.992
	$\varepsilon''(T) = 0.001746T^2 - 0.2474T + 18.41$	0.916
Pear	$\varepsilon'(T) = -0.0008304T^2 - 0.05229T + 71.06$	0.896
	$\varepsilon''(T) = 0.001453T^2 - 0.2498T + 20.95$	0.854

From: [19].

Table 3.39
Dielectric Properties (Mean ± STD of Two Replicates) of Fruits and Nuts

Sample	f (MHz)	T°C	20	30	40	50	60
Golden Delicious apple	915	ε'	74.3±0.8	72.3±0.7	70±0.8	67.8±1.0	65.6±1.0
		ε''	8.5±0.0	8.5±1.1	8.2±0.9	8.3±0.6	8.7±0.3
	1,800	ε'	67.4±0.9	66±0.9	64.1±0.9	62.1±1.0	60.1±1.0
		ε''	9.9±0.1	8.7±0.0	7.6±0.0	6.9±0.1	6.7±0.1
Red Delicious apple	915	ε'	77±0.0	74.5±0.2	71.5±0.1	68.9±0.2	67.1±0.5
		ε''	10.0±1.4	9.4±1.8	10.0±2.5	9.8±2.8	8.9±1.9
	1,800	ε'	70.4±0.5	68.3±0.4	66.1±0.5	64.0±0.5	62.0±0.8
		ε''	10.8±0.2	9.4±0.7	8.3±0.7	7.4±0.8	6.7±0.7
Almond	915	ε'	5.7±0.5	6.4±1.8	6.0±1.3	5.7±0.1	6.4±1.3
		ε''	1.7±0.9	3.2±2.3	3.3±2.0	3.4±0.5	3.1±1.4
	1,800	ε'	5.8±0.2	3.4±2.3	3.6±2.1	4.2±1.6	3.9±2.3
		ε''	2.9±0.8	3.4±0.9	3.5±0.7	3.4±0.2	3.0±1.2
Cherry	915	ε'	73.7±0.1	72.0±0.3	69.6±0.7	66.7±1.6	64.1±1.8
		ε''	16.4±0.0	17.2±0.5	18.3±1.0	19.3±1.4	20.4±1.9
	1,800	ε'	70.9±0.1	69.7±0.3	67.8±0.6	65.2±1.5	62.8±1.6
		ε''	16.0±0.2	15.1±0.6	14.6±0.9	14.2±1.1	14.1±1.4
Grapefruit	915	ε'	72.7±2.5	70.8±2.3	68.5±2.1	66.1±2.1	63.7±2.0
		ε''	12.1±0.0	12.5±0.2	13.3±0.4	14.2±0.3	15.5±0.3
	1,800	ε'	72.1±1.2	70.2±1.1	68.2±1.1	66.0±0.9	63.7±0.8
		ε''	12.6±0.1	11.5±0.2	10.9±0.2	10.7±0.2	10.7±0.2
Orange	915	ε'	72.9±1.9	70.6±1.8	68.0±2.1	66.1±0.6	63.2±0.7
		ε''	16.5±2.8	17.8±2.7	18.7±3.0	17.5±1.2	18.4±1.2
	1,800	ε'	72.5±0.1	70.7±0.3	68.6±0.4	65.6±0.2	62.7±0.3
		ε''	14.8±0.5	13.9±0.5	13.1±0.5	12.3±0.2	12.2±0.2
Walnut	915	ε'	2.2±1.6	2.1±0.3	3.0±0.1	3.4±0.0	3.8±0.0
		ε''	2.9±0.1	2.6±0.1	2.3±0.1	2.0±0.0	1.8±0.0
	1,800	ε'	2.1±0.7	2.7±0.2	3.2±0.0	3.5±0.0	3.7±0.0
		ε''	1.8±0.2	1.6±0.2	1.3±0.2	1.1±0.1	1.0±0.1

From: [20].

Thermal parameters of vegetables and fruits are well studied in [10, 12, 21]. For most of them three main parameters: heat capacity, thermal conductivity, and density are known as a function of temperature and moisture. This data for potato, beet, and carrot are shown in Tables 3.40 to 3.43. Some parameters in Table 3.41 are defined over n and q, where n is the volume fraction of dry matter and q is the moisture content (wet basis).

Table 3.40
Thermal Parameters of Fresh Potato with Moisture Content $0.723 \leq X_w \leq 0.836$ (Wet Basis)

Parameter	Temperature (°C)								
	25	40	50	60	70	80	90	105	130
λ_t, W/(m·K)	0.533	0.411	0.470	0.460	0.510	0.560	0.561	0.639	0.641
ρ_t, kg/m^3	1040	1127	1117	1121	1105	1107	1103	—	—
C_t, J/(kg·K)	6398	4196	4439	3842	3737	3477	2921	2536	—

From: [12].

Table 3.41
Thermal Parameters of Potato Food Products

Product	Temperature	Parameters	Units
Frozen potato garnish	$233 \leq T°K \leq 268$	$\rho_t(T) = 0.0011T^3 - 0.7834T^2 + 191.07T - 14{,}530;$ $R^2 = 0.9886$	kg/m^3
		$1{,}880 \leq C_t(T) \leq 15{,}400$	J/(kg·K)
		$\lambda_t(T) = -0.0004T^2 + 0.1874T - 20.819; R^2 = 0.8928$	W/(m·K)
Fresh potato	$T = 283°K$	$\rho_t(n) = 987 + 465n; 0.1 < n < 0.35$	kg/m^3
	$273 \leq T°K \leq 313$	$C_t(q,T) = 68 + 4119q + 4.2T - 4.2qT; 0 \leq q \leq 1$	J/(kg·K)
	$283 \leq T°K \leq 333$	$\lambda_t(T) = 0.00525T - 0.896$	W/(m·K)
	$293 \leq T°K \leq 343$	$a_t \cdot 10^8(T) = 0.0751T - 7.3$	m^2/s
Dried potato	$293 \leq T°K \leq 343$	$\lambda_t(T) = 0.0033T - 0.42$	W/(m·K)
	$273 \leq T°K \leq 313$	$C_t(T) = 68 + 4.2T$	J/(kg·K)
Sweet potato	$283 \leq T°K \leq 313$	$933 \leq \rho_t(T) \leq 1037$	kg/m^3
		$1675 \leq C_t(T) \leq 2790$	J/(kg·K)
	$303 \leq T°K \leq 363$	$a_t \cdot 10^8(T) = 13.3$	m^2/s
	$363 \leq T°K \leq 393$	$a_t \cdot 10^8(T) = 31.45 - 0.05T$	

From: [21].

Thermal properties of root crops like potato, beet, and carrot depend not only on temperature and moisture but also on porosity (gases content in cells) and type of vegetable. As well, the core part of the vegetable differs slightly from the cladding part. Defrosted root crops demonstrate almost the same heat capacity and thermal conductivity as the fresh ones. Lower values of main thermal parameters are observed for dried root crops in comparison with fresh vegetables (Tables 3.42 and 3.43).

Table 3.42
Thermal Parameters of Beet

Vegetable	Temperature	Parameter	Units
Sugar beet	$293 \leq T°K \leq 333$	$863 \leq \rho_t(T) \leq 1,258$	kg/m^3
	$293 \leq T°K \leq 353$	$C_t(T) = 3862 - 3,900n + 1.3T + 3.3nT; \; 0 \leq n \leq 1$	J/(kg·K)
	$278 \leq T°K \leq 333$	$\lambda_t(T) = 4 \cdot 10^{-5}T^2 - 0.0199T + 3.092; \; R^2 = 0.9989$ $\lambda_t(T) = 0.0022T - 0.1$	W/(m·K)
	$273 \leq T°K \leq 333$	$a_t \cdot 10^8(T) = 13.4 + 0.032(T - 273)$	m^2/s
Dried beet	$288 \leq T°K \leq 298$	$1,550 \leq \rho_t(T) \leq 1,650$ $\rho_t(n) = 992 + 418n; \; 0.2 < n < 0.4$	kg/m^3
	$293 \leq T°K \leq 353$	$C_t(T) = 4.6T - 38$	J/(kg·K)
	$300 \leq T°K \leq 328$	$0.13 \leq \lambda_t(T) \leq 0.36$ $\lambda_t(q) = 0.47q + 0.174; \; 0.7 < q < 0.82$	W/(m·K)
Beet marc	$293 \leq T°K \leq 353$	$C_t(T) = 17.08T - 355$ $\lambda_t(T) = 0.0056T - 1.176$ $a_t \cdot 10^8(T) = 0.268T - 40.7$	J/(kg·K) W/(m·K) m^2/s

From: [21].

Table 3.43
Thermal Parameters of Carrot

Vegetable	Temperature	Parameter	Units
Carrot	$273 \leq T°K \leq 373$	$1,022 \leq \rho_t(T) \leq 1,034$ $3,650 \leq C_t(T) \leq 3,950$	kg/m^3 J/(kg·K)
	$293 \leq T°K \leq 363$	$\lambda_t(T) = 0.181 + 0.00097T; \; W = 89\%$ $a_t \cdot 10^8(T) = 0.098T - 16.75$	W/(m·K) m^2/s
Dried carrot	$T = 297°K$	$\rho_t = 833$ $C_t = 3140$	kg/m^3 J/(kg·K)
	$268 \leq T°K \leq 300$	$C_t(T) = 7.9T - 1087$ $\lambda_t = 0.3$	W/(m·K)

From: [21].

Both root crops and solanaceae vegetables (Table 3.44) consist of many chemical substances: proteins, fats, carbohydrates, acids, starch, salts, and so forth. Volume fractions of these substances define their chemical and physical properties. Thermal parameters of some cooked and fresh vegetables are shown in Tables 3.45 and 3.46. For most of them only the deviation intervals in various temperature ranges are known, but Table 3.44 is the exception, where the mathematical expressions for density, thermal conductivity, and heat capacity of tomato juice are shown.

Table 3.44
Thermal Parameters of Tomato, Pepper, and Eggplant

Vegetable	Temperature	Parameter	Units
Tomato $88 \leq W\% \leq 95$	$273 \leq T°K \leq 293$	$990 \leq \rho_t(T) \leq 1,064$	kg/m^3
		$3,726 \leq C_t(T) \leq 4,019$	J/(kg·K)
		$0.46 \leq \lambda_t(T) \leq 0.53$	W/(m·K)
	$T = 343°K$	$3,730 \leq C_t(T) \leq 3,752$	J/(kg·K)
		$0.5 \leq \lambda_t(T) \leq 0.54$	W/(m·K)
Tomato juice	$292 \leq T°K \leq 403$	$\rho_t(T) = 1143 + 475q - 0.5T$ $0.07 < q < 0.37$	kg/m^3
	$273 \leq T°K \leq 373$	$\rho_t(T) = 1,161 + 440q - 0.53T$ $0.05 < q < 0.55$	
	$293 \leq T°K \leq 433$	$C_t(T) = 4,187 + 7.1326qT - 4,407q,$ $0 < q < 0.8$	J/(kg·K)
	$293 \leq T°K \leq 353$	$C_t(T) = 4,187 + 3T - 3,726q;$ $0 < q < 1$	
		$\lambda_t(T) = 0.26 + 0.0011T - 0.43q$ $0.05 < q < 0.4$	W/(m·K)
	$273 \leq T°K \leq 353$	$\lambda_t(T) = 0.114 + 0.00144T; q = 0.316$	
Pepper $90 \leq W\% \leq 93$	$273 \leq T°K \leq 293$	$1,000 \leq \rho_t(T) \leq 1,014$	kg/m^3
		$3,936 \leq C_t(T) \leq 3,960$	J/(kg·K)
		$0.42 \leq \lambda_t(T) \leq 0.52$	W/(m·K)
Eggplant $79 \leq W\% \leq 92$	$273 \leq T°K \leq 293$	$990 \leq \rho_t(T) \leq 1,026$	kg/m^3
		$3,930 \leq C_t(T) \leq 4,030$	J/(kg·K)
		$0.35 \leq \lambda_t(T) \leq 0.38$	W/(m·K)

From: [21].

Table 3.45
Thermal Parameters of Cooked Vegetables

Vegetable	Temperature	Parameter	Units
Boiled cabbage	$363 \leq T°K \leq 383$	$3,930 \leq C_t(T) \leq 4,103$ $0.9 \leq \lambda_t(T) \leq 1.3$	J/(kg·K) W/(m·K)
Stewed cabbage, $78 \leq W\% \leq 89$	$303 \leq T°K \leq 343$	$858 \leq \rho_t(T) \leq 892$ $0.12 \leq \lambda_t(T) \leq 0.42$	kg/m^3 W/(m·K)
Mashed pumpkin with rice	$293 \leq T°K \leq 353$	$\rho_t(T) = 1,348 - 0.467T$ $C_t(T) = 3,503 + 0.45T$	kg/m^3 J/(kg·K)
Mashed tomato	$293 \leq T°K \leq 353$	$1,089 \leq \rho_t(T) \leq 1,100$ $3,342 \leq C_t(T) \leq 3,720$ $\lambda_t(T) = 0.115 + 0.00144T$	kg/m^3 J/(kg·K) W/(m·K)
Navy beans	$299 \leq T°K \leq 395$	$a_t \cdot 10^8(T) = 18$	m^2/s

From: [21].

Table 3.46
Thermal Parameters of Different Fresh Vegetables

Vegetable	Temperature	Parameter	Units
Turnip, $85 \leq W\% \leq 92$	$273 \leq T°K \leq 321$	$940 \leq \rho_t(T) \leq 984$ $3,760 \leq C_t(T) \leq 3,890$ $0.48 \leq \lambda_t(T) \leq 0.54$	kg/m^3 $J/(kg \cdot K)$ $W/(m \cdot K)$
Celery, $W = 88\%$	$293 \leq T°K \leq 313$	$840 \leq \rho_t(T) \leq 1,002$ $3,665 \leq C_t(T) \leq 4,232$ $0.51 \leq \lambda_t(T) \leq 0.58$	kg/m^3 $J/(kg \cdot K)$ $W/(m \cdot K)$
Black radish, $W = 93\%$	$293 \leq T°K \leq 313$	$800 \leq \rho_t(T) \leq 950$ $C_t(q) = 1,388 + 2,799q$ $0 < q < 1$	kg/m^3 $J/(kg \cdot K)$
White cabbage, $91 \leq W\% \leq 97$	$273 \leq T°K \leq 293$	$400 \leq \rho_t(T) \leq 850$ $3,890 \leq C_t(T) \leq 3,970$ $0.3 \leq \lambda_t(T) \leq 0.4$	kg/m^3 $J/(kg \cdot K)$ $W/(m \cdot K)$
Lettuce, $W = 95\%$	$273 \leq T°K \leq 293$	$480 \leq \rho_t(T) \leq 550$ $3,978 \leq C_t(T) \leq 4,070$ $0.1 \leq \lambda_t(T) \leq 0.2$	kg/m^3 $J/(kg \cdot K)$ $W/(m \cdot K)$
Onion, $81 \leq W\% \leq 87$	$273 \leq T°K \leq 293$	$795 \leq \rho_t(T) \leq 1,010$ $2,638 \leq C_t(T) \leq 3,832$ $0.17 \leq \lambda_t(T) \leq 0.57$	kg/m^3 $J/(kg \cdot K)$ $W/(m \cdot K)$
Garlic, $62 \leq W\% \leq 64$	$273 \leq T°K \leq 293$	$936 \leq \rho_t(T) \leq 1,096$ $C_t(q) = 1,420 + 2,767q$ $0 < q < 1$ $0.4 \leq \lambda_t(T) \leq 0.6$	kg/m^3 $J/(kg \cdot K)$ $W/(m \cdot K)$
Pumpkin, $89 \leq W\% \leq 95$	$273 \leq T°K \leq 293$	$942 \leq \rho_t(T) \leq 861$ $3,850 \leq C_t(T) \leq 3,982$ $0.47 \leq \lambda_t(T) \leq 0.52$	kg/m^3 $J/(kg \cdot K)$ $W/(m \cdot K)$
Squash, $87 \leq W\% \leq 95$	$273 \leq T°K \leq 293$	$855 \leq \rho_t(T) \leq 917$ $3,580 \leq C_t(T) \leq 4,040$ $0.48 \leq \lambda_t(T) \leq 0.52$	kg/m^3 $J/(kg \cdot K)$ $W/(m \cdot K)$
Cucumber, $95 \leq W\% \leq 97$	$273 \leq T°K \leq 293$	$904 \leq \rho_t(T) \leq 1,019$ $4,057 \leq C_t(T) \leq 4,103$ $0.74 \leq \lambda_t(T) \leq 0.85$	kg/m^3 $J/(kg \cdot K)$ $W/(m \cdot K)$
Melon, $95 \leq W\% \leq 97$	$273 \leq T°K \leq 293$	$930 \leq \rho_t(T) \leq 1,022$ $3,840 \leq C_t(T) \leq 3,981$ $0.48 \leq \lambda_t(T) \leq 0.57$	kg/m^3 $J/(kg \cdot K)$ $W/(m \cdot K)$

From: [21].

Moisture is one of the most important factors influencing both the dielectric and thermal properties of fruit. For example, the CDP of apples at room temperature is expressed [22] as:

$$\varepsilon' = 1.9307W^2 + 1.7208W + 10.296 \qquad (3.4)$$

$$\varepsilon'' = 0.4585W^2 + 1.26W + 2.7 \qquad (3.5)$$

The thermal parameters of some fruits are listed in Tables 3.47 to 3.51.

Table 3.47
Thermal Properties of Pears

Product	Temperature	Parameter	Units
Fresh pears	$283 \leq T°K \leq 293$	$989 \leq \rho_t(T) \leq 1{,}144$, $0 < W\% < 87.5$	kg/m³
		$3{,}584 \leq C_t(T) \leq 3{,}810$, $76 < W\% < 86$	J/(kg·K)
	$296 \leq T°K \leq 303$	$0.595 \leq \lambda_t(T) \leq 0.92$, $W = 85\%$	W/(m·K)
Pear juice	$283 \leq T°K \leq 293$	$3{,}590 \leq C_t(T) \leq 3{,}610$, $W = 83.5\%$	J/(kg·K)
		$0.476 \leq \lambda_t(T) \leq 0.56$, $W = 85\%$	W/(m·K)

From: [21].

Table 3.48
Thermal Properties of Apples

Product	Temperature	Parameter	Units
Fresh apples	$273 \leq T°K \leq 293$	$780 \leq \rho_t(T) \leq 890$	kg/m³
	$293 \leq T°K \leq 363$	$840 \leq \rho_t(T) \leq 950$	
	$273 \leq T°K \leq 333$	$C_t(T) = 7.5T + 1{,}312$, $q = 0.127$	J/(kg·K)
		$C_t(T) = 4.9T + 2{,}062$, $q = 0.142$	
	$293 \leq T°K \leq 363$	$3{,}739 \leq C_t(T) \leq 3{,}820$	
		$0.41 \leq \lambda_t(T) \leq 0.5$	W/(m·K)
Dried apples	$273 \leq T°K \leq 293$	$\rho_t(q) = 1{,}520 - 522q$, $0 < q < 1$	kg/m³
		$C_t(q) = 1{,}220 + 6170q - 4{,}580q^2$, $0.24 < q < 0.76$	J/(kg·K)
		$\lambda_t(q) = 0.033 + 0.0565q + 0.411q^2$, $0.24 < q < 0.76$	W/(m·K)
Frozen apples	$233 \leq T°K \leq 272$	$799 \leq \rho_t(T) \leq 810$	kg/m³
	$258 < T°K \leq 263$	$C_t(T) = 0.76T - 182.6$, $q = 0.585$	J/(kg·K)
		$C_t(T) = 0.79T - 191.5$, $q = 0.873$	
	$263 < T°K \leq 272$	$C_t(T) = 390.3 - 1.42T$, $q = 0.585$	
		$C_t(T) = 406.2 - 1.48T$, $q = 0.873$	
	$248 < T°K \leq 272$	$\lambda_t(T) = 13.3q - 0.0103T - 7.488$, $0.856 < q < 0.873$	W/(m·K)
	$233 \leq T°K \leq 272$	$0.4 \leq \lambda_t(T) \leq 1.66$	

From: [21].

Table 3.49
Thermal Properties of Apple-Based Products

Product	Temperature	Parameter	Units
Apple pulp	$273 \leq T°K \leq 293$	$990 \leq \rho_t(T) \leq 1{,}058$	kg/m^3
		$3{,}881 \leq C_t(T) \leq 3{,}927$	$J/(kg \cdot K)$
	$298 \leq T°K \leq 333$	$\lambda_t(T) = 0.02T - 5.47$	$W/(m \cdot K)$
	$T = 302°K$	$0.4 \leq \lambda_t(T) \leq 0.6$	
Apple mash	$293 \leq T°K \leq 302$	$990 \leq \rho_t(T) \leq 1{,}050$	kg/m^3
		$3{,}420 \leq C_t(T) \leq 3{,}726$	$J/(kg \cdot K)$
		$0.513 \leq \lambda_t(T) \leq 0.692$	$W/(m \cdot K)$
Apple with apricot mash	$293 \leq T°K \leq 353$	$\rho_t(T) = 1{,}244 - 0.467T$	kg/m^3
		$C_t(T) = 3{,}211 + 0.251T$	$J/(kg \cdot K)$
Apple with cherry mash	$293 \leq T°K \leq 353$	$\rho_t(T) = 1{,}178 - 0.282T$	kg/m^3
		$C_t(T) = 2{,}026 + 4.3T$	$J/(kg \cdot K)$
Apple jam, $W = 32.5\%$	$273 \leq T°K \leq 293$	$1{,}294 \leq \rho_t(T) \leq 1{,}341$	kg/m^3
		$2{,}258 \leq C_t(T) \leq 2{,}262$	$J/(kg \cdot K)$
		$0.3 \leq \lambda_t(T) \leq 0.32$	$W/(m \cdot K)$
Apple syrup	$293 \leq T°K \leq 343$	$\rho_t(T) = 1{,}503 - 0.57T$	kg/m^3
		$C_t(T) = 1{,}044.2 ln(T) - 3{,}156.8$, $R^2 = 0.9743$	$J/(kg \cdot K)$
	$273 \leq T°K \leq 373$	$0.42 \leq \lambda_t(T) \leq 0.55$	$W/(m \cdot K)$
Apple juice	$273 \leq T°K \leq 363$	$\rho_t(T) = 1{,}084 + 600n - 0.308T - 0.407nT$, $0.12 < n < 0.74$	kg/m^3
	$293 \leq T°K \leq 363$	$C_t(T) = 4{,}187 - 3{,}526n + 3.6nT$, $0.1 < n < 0.7$	$J/(kg \cdot K)$
	$303 \leq T°K \leq 323$	$0.49 \leq \lambda_t(T) \leq 0.56$, $n = 0.21$	$W/(m \cdot K)$

From: [21].

Table 3.50
Thermal Properties of Peaches and Apricots

Product	Temperature	Parameter	Units
Fresh peaches	$273 \leq T°K \leq 300$	$950 \leq \rho_t(T) \leq 1{,}010$	kg/m^3
		$3{,}848 \leq C_t(T) \leq 3{,}936$, $88 < W\% < 92$	$J/(kg \cdot K)$
		$0.49 \leq \lambda_t(T) \leq 0.59$	$W/(m \cdot K)$
		$a_t(T) \cdot 10^8 = -9 + 0.085T$	m^2/s
Frozen peaches	$233 \leq T°K \leq 266$	$920 \leq \rho_t(T) \leq 970$	kg/m^3
		$1{,}717 \leq C_t(T) \leq 7{,}704$	$J/(kg \cdot K)$
Fresh apricots	$283 \leq T°K \leq 293$	$886 \leq \rho_t(T) \leq 1{,}109$	kg/m^3
		$2{,}500 \leq C_t(T) \leq 3{,}768$	$J/(kg \cdot K)$
		$0.475 \leq \lambda_t(T) \leq 0.525$	$W/(m \cdot K)$

From: [21].

Table 3.51
Thermal Properties of Oranges and Orange-Based Products

Product	Temperature	Parameter	Units
Fresh orange	$283 \leq T°K \leq 293$	$791 \leq \rho_t(T) \leq 919$, $80 < W\% < 85$	kg/m^3
	$283 \leq T°K \leq 293$	$2{,}937 \leq C_t(T) \leq 3{,}920$, $58 < W\% < 89$	$J/(kg \cdot K)$
	$T = 293°$	$C_t(q) = 1{,}520 + 2{,}667q$, $0 < q < 1$	
	$293 \leq T°K \leq 313$	$0.41 \leq \lambda_t(T) \leq 0.58$, $80 < W\% < 85$	$W/(m \cdot K)$
Orange juice	$283 \leq T°K \leq 293$	$\rho_t(q) = 1{,}418 - 420q$, $0 < q < 1$	kg/m^3
	$288 \leq T°K \leq 305$	$3{,}680 \leq C_t(T) \leq 4{,}150$, $W = 89\%$	$J/(kg \cdot K)$
	$293 \leq T°K \leq 353$	$0.559 \leq \lambda_t(T) \leq 0.631$, $W = 87.4\%$	$W/(m \cdot K)$
		$0.389 \leq \lambda_t(T) \leq 0.436$, $W = 36\%$	
Orange syrup	$293 \leq T°K \leq 343$	$\rho_t(T) = 1{,}549 - 0.67T$	kg/m^3
		$C_t(T) = 1{,}820 + 3.52T$	$J/(kg \cdot K)$
	$303 \leq T°K \leq 343$	$0.406 \leq \lambda_t(T) \leq 0.563$	$W/(m \cdot K)$

From: [21].

3.6 Berries, Mushrooms, and Beans

Information about the physical properties of berries, mushrooms, and beans is quite limited. A few examples of experimental studies from the literature are represented in Tables 3.52 to 3.58. The dielectric measurements were performed in [23] with a cavity perturbation technique. Interpolation expressions for CDP of different berries, fruits, and mushrooms (about 150 samples) as a function of temperature, moisture, and density have also been derived from [23].

Table 3.52
Dielectric Properties of Some Agricultural Products at 2,800 MHz

Agro Product	W%	$T = 20°C$		$T = 40°C$		$T = 60°C$	
		ε'	ε''	ε'	ε''	ε'	ε''
Mushroom	72.2	64 ± 0.1	27 ± 0.8	54 ± 1.4	21 ± 0.5	55 ± 0.6	23 ± 0.4
	38.65	31 ± 6.5	20 ± 3.0	39 ± 7.0	22 ± 2.5	44 ± 7.7	20 ± 4.5
Apple	75	48 ± 1.5	16 ± 0.5	48 ± 1.7	13 ± 0.8	48 ± 3.0	11 ± 1.4
	50	18 ± 5.0	9.5± 2.5	21 ± 5.0	10 ± 2.4	26 ± 7.0	10 ± 1.5
Parsley	88	62 ± 0.7	18 ± 1.6	49 ± 0.1	17 ± 0.2	45 ± 1.1	18 ± 0.5
	46.8	36 ± 2.2	21 ± 1.6	37 ± 1.8	23 ± 1.1	43 ± 5.9	26 ± 0.8
Chervil	93	63 ± 0.3	18 ± 1	63 ± 0.2	18 ± 1	62 ± 0.3	19 ± 1
	57.9	40 ± 8.7	26 ± 1.9	43 ± 6.4	27 ± 0.7	38 ± 8.5	28 ± 0.5
Strawberry	79.1	62 ± 4.9	18 ± 0.7	62 ± 6.6	15 ± 0.4	60 ± 4.4	16 ± 0.5
	50	33 ± 9.4	17 ± 0.4	43 ± 8.7	16 ± 0.4	43 ± 7.7	15 ± 0.8

From: [23].

Table 3.53
Dielectric Properties of Some Berries
at 2,450 MHz and $T = 20°C$

Type of Berry	W%	ε'	ε''
Raspberry	73.4	54	13.4
Black currant	76.3	56.2	14.2
	79.9	58	14.3
Gooseberry	80	58.1	14.5

From: [3].

Table 3.54
Thermal Properties of Coffee Beans and Cocoa Beans

Beans	Temperature	Parameters	Unit
Coffee, W = 12%	$20 \leq T°C \leq 140$	$C_t(T) = -2 \cdot 10^{-4}T^3 + 0.0149T^2 + 6.2896T + 1257.9$; $R^2 = 0.9968$	$J/(kg \cdot K)$
		$\lambda_t(T) = -10^{-5}T^2 + 0.0026T + 0.3329$; $R^2 = 0.9963$	$W/(m \cdot K)$
		$\rho_t = 1,450$	kg/m^3
Cocoa	$20 \leq T°C \leq 110$	$C_t = 2,261$	$J/(kg \cdot K)$
		$\lambda_t(T) = 0.105 \div 0.093$	$W/(m \cdot K)$
		$\rho_t = 560$	kg/m^3

From: [10].

Table 3.55
Thermal Characteristics of Beans and Peas

Vegetable	Temperature	Parameter	Units
Navy beans, $10 \leq W\% \leq 13$	$273 \leq T°K \leq 293$	$1,180 \leq \rho_t(T) \leq 1,360$	kg/m^3
	$303 \leq T°K \leq 343$	$C_t(q) = 1,480 + 2,710q,\ 0.1 < q < 0.35$	J/(kg·K)
	$283 \leq T°K \leq 295$	$0.139 \leq \lambda_t(T) \leq 0.174$	W/(m·K)
Green peas	$273 \leq T°K \leq 293$	$1,110 \leq \rho_t(T) \leq 1,370$	kg/m^3
	$303 \leq T°K \leq 343$	$C_t(q) = 1,232 + 4,650q,\ 0.05 < q < 0.14$	J/(kg·K)
	$273 \leq T°K \leq 373$	$1,620 \leq C_t(T) \leq 3,684$	
	$273 \leq T°K \leq 293$	$\lambda_t(T) = 0.00224T - 0.2425,\ q = 0.745$	W/(m·K)
	$276 \leq T°K \leq 290$	$0.18 \leq \lambda_t(T) \leq 0.3,\ q < 0.15$	
Beans	$273 \leq T°K \leq 293$	$1,215 \leq \rho_t(T) \leq 1,310$	kg/m^3
	$303 \leq T°K \leq 338$	$C_t(T) = 7.1T - 351,\ q = 0.115$	J/(kg·K)
	$318 \leq T°K \leq 338$	$C_t(T) = -951 + 5,138q + 8.06T - 8.06Tq,\ 0 < q < 1$	
	$273 \leq T°K \leq 302$	$0.356 \leq \lambda_t(T) \leq 0.7$	W/(m·K)

From: [21].

It is well known that mushrooms are very porous products. Fresh mushrooms contain up to 30% air. When pores are filled with water, the density and thermal conductivity of mushrooms are increased (Table 3.56). The porosity of fresh cherries is much less (about 2%) and moisture content is much higher. Cherries are a type of berry that contains pits. The physical properties of cherries (Table 3.57) in the literature are usually given without taking into account the pit parameters. Additional data on the density, heat capacity, and thermal conductivity of various berries are represented in Table 3.58. The deviation of the thermal parameters of berries in accordance with geographical position is not essential [21]. The thermal conductivity of frozen grape juice is higher at low temperatures because of the crystallization process. A large variety of mushrooms and berries makes derivation of universal mathematical equations predicting their physical properties very difficult.

Table 3.56
Thermal Parameters of Mushrooms

Product	Temperature	Parameter	Units
Fresh mushrooms	$273 \leq T°K \leq 293$	$840 \leq \rho_t(T) \leq 960$	kg/m^3
	$273 \leq T°K \leq 373$	$3,840 \leq C_t(T) \leq 3,940$	$J/(kg \cdot K)$
	$273 \leq T°K \leq 293$	$0.23 \leq \lambda_t(T) \leq 0.49$	$W/(m \cdot K)$
Dried mushrooms	$T = 293°K$	$\rho_t(n) = 1,000 + 517n; 0 < n < 1$	kg/m^3
	$303 \leq T°K \leq 373$	$C_t(T) = 0.64T + 1376; q = 0.01$	$J/(kg \cdot K)$
	$273 \leq T°K \leq 293$	$0.01 \leq \lambda_t(T) \leq 0.03$	$W/(m \cdot K)$
Boiled mushrooms	$290 \leq T°K \leq 293$	$1,000 \leq \rho_t(T) \leq 1,100$	kg/m^3
		$3,580 \leq C_t(T) \leq 3,780$	$J/(kg \cdot K)$
		$0.58 \leq \lambda_t(T) \leq 0.64$	$W/(m \cdot K)$

From: [21]

Table 3.57
Thermal Properties of Cherry and Cherry-Based Products

Product	Temperature	Parameter	Units
Fresh cherries	$283 \leq T°K \leq 293$	$1,053 \leq \rho_t(T) \leq 1,083; 80 < W\% < 85$	kg/m^3
		$3,349 \leq C_t(T) \leq 3,850; 71 < W\% < 82$	$J/(kg \cdot K)$
		$0.465 \leq \lambda_t(T) \leq 0.515$	$W/(m \cdot K)$
	$298 \leq T°K \leq 348$	$a_t(T) \cdot 10^8 = (5T - 628)^{-1}$	m^2/s
Cherry juice	$293 \leq T°K \leq 353$	$1,011 \leq \rho_t(T) \leq 1,285; 0.1 < n < 0.6$	kg/m^3
	$283 \leq T°K \leq 293$	$3,490 \leq C_t(T) \leq 3,808; W = 86.7\%$	$J/(kg \cdot K)$
		$0.52 \leq \lambda_t(T) \leq 0.57, W = 86.7\%$	$W/(m \cdot K)$

From: [21].

Table 3.58
Thermal Properties of Berries

Product	Temperature	Parameter	Units
Fresh grapes	$283 \leq T°K \leq 293$	$1{,}070 \leq \rho_t(T) \leq 1{,}137$ $3{,}559 \leq C_t(T) \leq 3{,}840$; $74 < W\% < 84$ $0.49 \leq \lambda_t(T) \leq 0.52$; $79 < W\% < 85$	kg/m^3 $J/(kg \cdot K)$ $W/(m \cdot K)$
Grape juice	$273 \leq T°K \leq 343$	$\rho_t(T) = 1{,}037 + 686n - 0.25T - 0.42nT,$ $0.15 < n < 0.75$	kg/m^3
	$293 \leq T°K \leq 353$	$2{,}901 \leq C_t(T) \leq 3{,}718$	$J/(kg \cdot K)$
	$273 \leq T°K \leq 333$	$0.509 \leq \lambda_t(T) \leq 0.626$	$W/(m \cdot K)$
Frozen grape juice	$233 \leq T°K \leq 273$	$0.28 \leq \lambda_t(T) \leq 0.952$; $0.1 < n < 0.4$	$W/(m \cdot K)$
	$233 \leq T°K \leq 268$	$a_t(T) \cdot 10^8 = 94.1 - 0.349T$; $n = 0.3$	m^2/s
Red currant	$283 \leq T°K \leq 293$	$1{,}016 \leq \rho_t(T) \leq 1{,}077$ $3{,}609 \leq C_t(T) \leq 3{,}684$; $81 < W\% < 87$	kg/m^3 $J/(kg \cdot K)$
Cranberry	$283 \leq T°K \leq 293$	$1{,}060 \leq \rho_t(T) \leq 1{,}064$ $3{,}810 \leq C_t(T) \leq 3{,}831$; $87 < W\% < 89$	kg/m^3 $J/(kg \cdot K)$
Raspberry	$283 \leq T°K \leq 293$	$950 \leq \rho_t(T) \leq 1{,}030$; $80 < W\% < 82$ $3{,}475 \leq C_t(T) \leq 3{,}750$; $79 < W\% < 84$ $0.49 \leq \lambda_t(T) \leq 0.51$	kg/m^3 $J/(kg \cdot K)$ $W/(m \cdot K)$

From: [21].

3.7 Oils

The dielectric properties of corn oil at 3 GHz in temperature range $10 \leq T°C \leq 60$ [24, 25] are:

$$\varepsilon'(T) = 3.998 \cdot 10^{-6} T^2 + 0.001313T + 2.587 \qquad (3.6)$$

$$\varepsilon''(T) = 1.454 \cdot 10^{-6} T^2 + 0.00172T + 0.1214 \qquad (3.7)$$

Available information about various oils shows that both the dielectric permittivity and loss factor are quite low at ISM bands. This property is used often in scientific investigations of electromagnetic and temperature fields in microwave heating systems [25].

The thermal parameters of food oils are more widely studied than many other foods. In Table 3.59 one can see mathematical expressions (linear functions) describing temperature dependencies of density, thermal conductivity, and heat capacity of oils, taken from [10]. Pressure does not influence significantly on the density of sunflower oil during technological processing [10]. The heat capacity of refined oil is less than that of unrefined oil (see Table 3.59).

Table 3.59
Thermal Properties of Food Oils

Oil	Temperature	Parameters	Units
Refined sunflower oil	$293 \leq T°K \leq 458$	$\rho_t(T) = 1,107 - 0.6175T$	kg/m³
	$313 \leq T°K \leq 413$	$C_t(T) = 551 + 4.19T$	J/(kg·K)
Unrefined sunflower oil	$293 \leq T°K \leq 458$	$\rho_t(T) = 1,098 - 0.605T$	kg/m³
	$253 \leq T°K \leq 413$	$\rho_t(T) = 1115.7 - 0.68T$	
	$293 \leq T°K \leq 373$	$C_t(T) = 1,031 + 3.025T$	J/(kg·K)
	$298 \leq T°K \leq 343$	$\lambda_t(T) = 0.3397 - 0.00057T$	W/(m·K)
Refined cottonseed oil	$263 \leq T°K \leq 423$	$C_t(T) = 500 + 4.21T$	J/(kg·K)
	$273 \leq T°K \leq 413$	$\lambda_t(T) = 0.207 + 0.000136T$	W/(m·K)
Unrefined cottonseed oil	$253 \leq T°K \leq 413$	$\rho_t(T) = 1120.6 - 0.68T$	kg/m³
	$298 \leq T°K \leq 353$	$\rho_t(T) = 1,106 - 0.62T$	
	$313 \leq T°K \leq 423$	$C_t(T) = 507 + 4.26T$	J/(kg·K)
	$173 \leq T°K \leq 218$	$C_t(T) = 7.2T - 115$	
Refined olive oil	$293 \leq T°K \leq 458$	$\rho_t(T) = 1098.8 - 0.636T$	kg/m³
	$283 \leq T°K \leq 323$	$C_t(T) = 1,639 + 1.05T$	J/(kg·K)
	$293 \leq T°K \leq 373$	$\lambda_t(T) = 0.178 - 0.0000375T$	W/(m·K)
Unrefined olive oil	$293 \leq T°K \leq 458$	$\rho_t(T) = 1108.6 - 0.6725T$	kg/m³
	$283 \leq T°K \leq 333$	$\rho_t(T) = 1,110 - 0.678T$	
Corn oil	$253 \leq T°K \leq 413$	$\rho_t(T) = 1,119 - 0.68T$	kg/m³
	$283 \leq T°K \leq 333$	$\rho_t(T) = 1,130 - 0.701T$	
	$323 \leq T°K \leq 423$	$C_t(T) = 795 + 3.14T$	J/(kg·K)
	$253 \leq T°K \leq 413$	$\lambda_t(T) = 0.192 \cdot 10^{-4}(1,130 - 0.701T)^{4/3}$	W/(m·K)
Soy oil	$293 \leq T°K \leq 458$	$\rho_t(T) = 1,115 - 0.633T$	kg/m³
	$273 \leq T°K \leq 353$	$C_t(T) = 1,149 + 2.65T$	J/(kg·K)
	$353 \leq T°K \leq 483$	$C_t(T) = 957 + 3.1T$	
	$253 \leq T°K \leq 458$	$\lambda_t(T) = 0.194 \cdot 10^{-4}(1119.6 - 0.68T)^{4/3}$	W/(m·K)
Sesame oil	$253 \leq T°K \leq 413$	$\rho_t(T) = 1117.7 - 0.68T$	kg/m³
	$283 \leq T°K \leq 333$	$\rho_t(T) = 1,120 - 0.68T$	
	$273 \leq T°K \leq 353$	$C_t(T) = 1051 + 3.04T$	J/(kg·K)
	$363 \leq T°K \leq 413$	$C_t(T) = 935 + 3.35T$	
	$253 \leq T°K \leq 413$	$\lambda_t(T) = 0.1985 \cdot 10^{-4}(1117.7 - 0.68T)^{4/3}$	W/(m·K)

From: [10].

3.8 Some Thermal Parameters of Different Foods

The thermal properties of some foods are listed in Tables 3.60–3.63. The essential inhomogeneity of these products in most cases again does not allow us to derive predictive mathematical equations of the thermal parameters. Sometimes information about separate food components helps to obtain such equations (Tables 3.64 and 3.65). The important parameter of the viscosity of liquid foods is shown in Table 3.66.

Table 3.60
Heat Capacity C_t [kJ/(kg·K)] of Frozen Solid and Liquid Food Products

$T°K$	233	243	253	263	265	267	269	271	272
Fish	1.8	2	2.5	4.1	5.3	7.7	15.3	67.4	108.6
Meat	1.93	1.93	2.47	4.6	5.02	8.16	15.9	77.8	233.9
Orange juice	1.9	2.09	2.55	5.19	7.16	11.4	23.5	40.6	89.16

From: [12].

Table 3.61
Thermal Diffusivity of Different Foods

Food	$a_t \cdot 10^7$ m²/s	Temperature	Moisture Content, Wet Basis
Meat gravy	1.46	$60 \leq T°C \leq 112$	0.773
Meat croquette	1.98	$59 \leq T°C \leq 115$	0.74
Cooked seaweed	1.9	$56 \leq T°C \leq 110$	0.76
Frankfurters	2.36	$58 \leq T°C \leq 109$	0.734
Cooked lentil and rice	1.98	$60 \leq T°C \leq 114$	0.8
Chicken and rice	1.93	$65 \leq T°C \leq 113$	0.75
Chicken, potato, and rice	1.7	$72 \leq T°C \leq 109$	0.737
Beans, corn, and squash	1.83	$56 \leq T°C \leq 113$	0.683
Meat, tomato, and potato	1.57	$65 \leq T°C \leq 106$	0.664
Meat, potato, and carrots	1.77	$56 \leq T°C \leq 113$	0.821

From: [12].

Table 3.62
Thermal Parameters of Candy Food Products

Product	Temperature	Parameters	Units
Candied fruit jelly	$298 \leq T°K \leq 358$	$1,361 \leq \rho_t(T) \leq 1,411$ $2,784 \leq C_t(T) \leq 3,500$ $0.25 \leq \lambda_t(T) \leq 0.39$	kg/m^3 J/(kg·K) W/(m·K)
Chocolate	$273 \leq T°K \leq 343$	$1,234 \leq \rho_t(T) \leq 1,320$ $1,482 \leq C_t(T) \leq 2,123$ $0.214 \leq \lambda_t(T) \leq 0.29$	kg/m^3 J/(kg·K) W/(m·K)
Monocrystal sugar	$273 \leq T°K \leq 288$ $273 \leq T°K \leq 363$ $293 \leq T°K \leq 353$ $273 \leq T°K \leq 363$	$\rho_t(T) = 1638 - 0.17T$ $C_t(T) = 4T + 58$ $C_t(T) = 5.44T - 396$ $\lambda_t(T) = 1.04 - 0.00178T$	kg/m^3 J/(kg·K) W/(m·K)
Amorphous sugar	$288 \leq T°K \leq 293$ $295 \leq T°K \leq 298$	$1,508 \leq \rho_t(T) \leq 1,554$ $C_t(T) = 176 + 4.19T$	kg/m^3 J/(kg·K)
Sugar sand	$284 \leq T°K \leq 303$	$879 \leq \rho_t(T) \leq 990$ $712 \leq C_t(T) \leq 1,365$ $0.13 \leq \lambda_t(T) \leq 0.4$	kg/m^3 J/(kg·K) W/(m·K)

From: [10].

Table 3.63
Thermal Parameters of Bread and Pastries

Product	Temperature	Parameters	Units
Bread crust, W = 5%	$284 \leq T°K \leq 387$	$417 \leq \rho_t(T) \leq 476$ $1,675 \leq C_t(T) \leq 2,050$ $0.056 \leq \lambda_t(T) \leq 0.195$	kg/m^3 J/(kg·K) W/(m·K)
Bread crumbs, $40 \leq W\% \leq 55$	$289 \leq T°K \leq 302$	$340 \leq \rho_t(T) \leq 690$ $2,448 \leq C_t(T) \leq 3,428$ $0.211 \leq \lambda_t(T) \leq 0.398$	kg/m^3 J/(kg·K) W/(m·K)
Cooked macaroni	$283 \leq T°K \leq 298$	$740 \leq \rho_t(T) \leq 760$ $3,870 \leq C_t(T) \leq 3,912$ $0.3 \leq \lambda_t(T) \leq 0.405$	kg/m^3 J/(kg·K) W/(m·K)
Biscuit dough, W = 37%	$293 \leq T°K \leq 299$ $293 \leq T°K \leq 373$	$1,024 \leq \rho_t(T) \leq 1,034$ $0.12 \leq \lambda_t(T) \leq 0.2$	kg/m^3 W/(m·K)
Wafer sheet	$288 \leq T°K \leq 358$	$1,040 \leq \rho_t(T) \leq 1,154$ $C_t(T) = 1088 + 1.38T$ $\lambda_t(T) = 0.0017T - 0.0054$	kg/m^3 J/(kg·K) W/(m·K)
Baked cookies	$298 \leq T°K \leq 358$	$520 \leq \rho_t(T) \leq 700$ $1,624 \leq C_t(T) \leq 2,181$ $0.1 \leq \lambda_t(T) \leq 0.283$	kg/m^3 J/(kg·K) W/(m·K)
Cookie dough	$288 \leq T°K \leq 313$	$1,280 \leq \rho_t(T) \leq 1,300$ $2,490 \leq C_t(T) \leq 2,530$ $0.31 \leq \lambda_t(T) \leq 0.35$	kg/m^3 J/(kg·K) W/(m·K)
Cake dough	$288 \leq T°K \leq 313$	$1,325 \leq \rho_t(T) \leq 1,340$ $2,659 \leq C_t(T) \leq 3,182$ $0.385 \leq \lambda_t(T) \leq 0.43$	kg/m^3 J/(kg·K) W/(m·K)

From: [10].

Table 3.64

Thermal Parameters of Some Components of Food Products

Component	Parameters	Units	Temperature
Protein	$\rho_t(T) = 1.33 \cdot 10^3 - 0.5184T$	kg/m^3	$0 \leq T°C \leq 150$
	$\lambda_t(T) = 0.179 + 0.0012T - 2.72 \cdot 10^{-6}T^2$	W/(m·K)	
	$C_t(T) = 2.0082 + 1.2089 \cdot 10^{-3}T - 1.3129 \cdot 10^{-6}T^2$	kJ/(kg·K)	
Fat	$\rho_t(T) = 9.2559 \cdot 10^2 - 0.41757T$	kg/m^3	$0 \leq T°C \leq 150$
	$\lambda_t(T) = 0.181 - 0.00276T - 1.77 \cdot 10^{-7}T^2$	W/(m·K)	
	$C_t(T) = 1.9842 + 1.4733 \cdot 10^{-3}T - 4.8008 \cdot 10^{-6}T^2$	kJ/(kg·K)	
Gelatin	$\lambda_t(T) = 0.303 + 0.0012T - 2.72 \cdot 10^{-6}T^2$	W/(m·K)	$0 \leq T°C \leq 100$

From: [12].

Table 3.65

Density of Some Components of Food Products at $T = 20°C$

Component	Density (kg/m^3)	Component	Density (kg/m^3)
Glucose	1,544	Protein	1400
Fructose	1,669	Fat	900 ÷ 950
Maltose	1,540	NaCl	2163
Gelatin	1,270	KCl	1988
Cellulose	1,270 ÷ 1,610	Glycerol	1260

From: [12].

Table 3.66
Specific Density and Kinematical Viscosity of Some Liquids

Liquid	T °C	Density (g/cm³)	T °C	Viscosity (10^{-6} m²/c)
Beer	15.6	1.01	20	1.8
Coconut oil	15.6	0.925	37.8	29.8 ÷ 31.6
			54.4	14.7 ÷ 15.7
Cod oil	15.6	0.928	37.8	32.1
			54.4	19.4
Corn oil	15.6	0.924	54.4	28.7
			100	8.6
Lard	15.6	0.96	37.8	62.1
			54.4	34.3
Lard oil	15.6	0.91 ÷ 0.93	37.8	41 ÷ 48
			54.4	23.4 ÷ 27.1
Olive oil	15.6	0.91 ÷ 0.92	37.8	43.2
			54.4	24.1
Peanut oil	15.6	0.92	37.8	42
			54.4	23.4
Soybean oil	15.6	0.924 ÷ 0.928	37.8	35.4
			54.4	19.64
Water, fresh	15.6	1.0	15.6	1.13
			54.4	0.55
Whale oil	15.6	0.925	37.8	35 ÷ 39
			54.4	20 ÷ 23.4

Finally, the temperature dependencies of the thermal conductivity and heat capacity of different foods and their components (taken from [26]) are given in Tables 3.67 and 3.68.

Table 3.67

Thermal Conductivity of Some Foods in
Temperature Range $0 \leq T°C \leq 100$

Food	Thermal Conductivity [W/m·K]
Water	$\lambda_t(T) = 0.57109 + 1.7625 \cdot 10^{-3}T$
Albumin	$\lambda_t(T) = 0.18068 + 1.1462 \cdot 10^{-3}T - 2.6888 \cdot 10^{-6}T^2$
Casein	$\lambda_t(T) = 0.17138 + 1.1234 \cdot 10^{-3}T - 2.4592 \cdot 10^{-6}T^2$
Whey protein	$\lambda_t(T) = 0.18627 + 1.2444 \cdot 10^{-3}T - 2.9499 \cdot 10^{-6}T^2$
Gluten	$\lambda_t(T) = 0.18671 + 1.3229 \cdot 10^{-3}T - 3.4197 \cdot 10^{-6}T^2$
Milk fat	$\lambda_t(T) = 0.17809 - 2.4381 \cdot 10^{-4}T - 5.5169 \cdot 10^{-7}T^2$
Vegetable oil	$\lambda_t(T) = 0.18224 - 2.1949 \cdot 10^{-4}T - 7.3411 \cdot 10^{-7}T^2$
Corn oil	$\lambda_t(T) = 0.18109 - 2.0145 \cdot 10^{-4}T - 7.8395 \cdot 10^{-7}T^2$
Lactose	$\lambda_t(T) = 0.1989 + 1.476 \cdot 10^{-3}T - 4.566 \cdot 10^{-6}T^2$
Sugar	$\lambda_t(T) = 0.20456 + 1.3744 \cdot 10^{-3}T - 4.2079 \cdot 10^{-6}T^2$
Cellulose	$\lambda_t(T) = 0.17944 + 1.3698 \cdot 10^{-3}T - 3.2086 \cdot 10^{-6}T^2$
Pectin	$\lambda_t(T) = 0.18644 + 1.2914 \cdot 10^{-3}T - 3.1286 \cdot 10^{-6}T^2$

Table 3.68

Heat Capacity of Some Foods

Food	Heat Capacity in [kJ/kg·K]	Temperature °C
Apple, Granny Smith	$C_t(T) = 3.4 + 0.0049T$	$0 \leq T \leq 60$
Apple, Golden Delicious	$C_t(T) = 3.36 + 0.0075T$	
Coffee, Mexican powdered	$C_t(T) = 0.921 + 0.007554T$	$45 \leq T \leq 150$
Coffee, Columbian powdered	$C_t(T) = 1.273 + 0.00646T$	
Peanut oil, hydrogenated	$C_t(T) = 1.97 + 0.00489T$	$46.8 \leq T \leq 76.8$
Peanut oil, unhydrogenated	$C_t(T) = 2.057 + 0.00167T$	$27 \leq T \leq 57$
Potato mix, fried	$C_t(T) = 3.0363 + 0.005951T$ $C_t(T) = 2.4036 + 0.00506T$	$50 \leq T \leq 100$ $T > 100$
Dried cornstarch	$C_t(T) = 1.0185 + 0.007789T$	$30 \leq T \leq 90$
Milk	$C_t(T,X_w) = 4.187 X_w$ $+ (1.373 + 0.0113T)(1 - X_w)$	$30 \leq T \leq 90$ $0.6 \leq X_w \leq 0.9$
Mushrooms, X_w–moisture content	$C_t(T,X_w) = 1.0217$ $+ 0.0092T + 2.47X_w$	$40 \leq T \leq 70$ $0.1 \leq X_w \leq 0.9$

Additional information about the temperature-dependent dielectric properties of foods such as tropical fruits, salmon fillets, chickpea flour, white bread, and eggs at 915 and 1.8 GHz is represented in [27–31].

References

[1]   Bengtsson, N. E., and T. Ohlsson, "Microwave Heating in the Food Industry," *Proceedings IEEE*, Vol. 62, No. 1, 1974, pp. 44–55.

[2]   Yang, X. H., and J. Tang, Eds., *Advances in Bioprocessing Engineering*, Hackensack, NJ: World Scientific Publishing, 2002.

[3]   Rogov, I. A., and S.V. Nekrytman, *Microwave Heating of Food Products*, Moscow: Agropromizdat, 1986 (in Russian).

[4]   Buffler, C. R., "Microwave Cooking and Processing," *Engineering Fundamentals for the Food Scientist*, New York: Van Nostrand Reinhold, 1993.

[5]   Decareau, R. V., *Microwaves in the Food Processing Industry*, Orlando, FL: Academic Press, Inc., 1985.

[6]   Metaxas, A. C., and R.J. Meredith, *Industrial Microwave Heating*, London: Peter Peregrinus, 1983.

[7]   Sipahioglu, O., S. A. Barringer, and C. Bircan, "The Dielectric Properties of Meat as a Function of Temperature and Composition," *Int. J. Microwave Power and Electromagnetic Energy*, 2003, Vol. 38, No. 3, pp. 161–169.

[8]   Barringer, S. A., O. Sipahioglu, and C. Bircan, "The Dielectric Properties of Fruits, Vegetables and Meat," *Microwave and Radio Frequency Applications*, The American Ceramic Society, 2003, pp. 57–67.

[9]   Gunasekaran, N., P. Mallikarjunan, and J. Eifert, et al., "Effect of Fat Content and Temperature on Dielectric Properties of Ground Beef," *Transactions ASAE*, Vol. 48, No. 2, 2005, pp. 673–680.

[10] Ginzburg, A. C., M. A. Gromov, and G. I. Krossovskaya, *Thermo Physical Characteristics of Food Products*, Moscow: Agropromizdat, 1990 (in Russian).

[11] Al-Holy, M., Y. Wang, and J. Tang, et al., "Dielectric Properties of Salmon (Oncorhynchus keta) and Sturgeon (Acipenser transmontanus) Caviar at Radio Frequency (RF) and Microwave (MW) Pasteurization Frequencies," *Journal of Food Engineering*, 2005, Vol.70, pp. 564–570.

[12] Rahman, S., *Food Properties Handbook*, London: CRC Press, 1995.

[13] Kudra, T., V. Raghavan, and C. Akyel, et al., "Electromagnetic Properties of Milk and Its Constituents at 2.45 GHz," *Int. J. Microwave Power and Electromagnetic Energy*, Vol. 27, No. 4, 1992, pp. 119–204.

[14] Herve, A. G., J. Tang, and L. Luedecke, et al., "Dielectric Properties of Cottage Cheese and Surface Treatment Using Microwaves," *Journal of Food Engineering*, Vol. 37, 1998, pp. 389–410.

[15] Kuchma, T. N., *Combined Methods of Microwave Pasteurisation of Some Food Products*, Ph.D. dissertation, Moscow Technological Institute of Food Industry, 1988 (in Russian).

[16] Wang, Y., T. D. Wig, and J. Tang, et al., "Dielectric Properties of Foods Relevant to RF and Microwave Pasteurization and Sterilization," *Journal of Food Engineering*, Vol. 57, 2003, pp. 257–268.

[17] Chen, H., J. Tang, and F. Liu, "Coupled Simulation of an Electromagnetic Heating Process Using the Finite Difference Time Domain Method," *Int. J. Microwave Power and Electromagnetic Energy*, Vol. 41, No. 3, 2007, pp. 50–68.

[18] Guan, D., M. Cheng, and Y. Wang, et al., "Dielectric Properties of Mashed Potatoes Relevant to Microwave and Radio-Frequency Pasteurization and Sterilization Processes," *Journal of Food Science*, Vol. 69, No. 1, 2004, pp. 30–37.

[19] Sipahioglu, O., and S. A. Barringer, "Dielectric Properties of Vegetables and Fruits as a Function of Temperature, Ash and Moisture Content," *Journal of Food Science*, Vol. 68, No. 1, 2003, pp. 234–239.

[20] Wang, S., J. Tang, and J. A. Johnson, et al., "Dielectric Properties of Fruits and Insect Pests as Related to Radio Frequency and Microwave Treatments," *Biosystems Engineering*, Vol. 85, No. 2, 2003, pp. 201–212.

[21] Ginzburg, A. C., and M. A. Gromov, *Thermo Physical Characteristics of Potatoes, Vegetables and Fruits*, Moscow: Agropromizdat, 1987 (in Russian).

[22] Diaz A., et al. "Mathematical Model of Combined Hot Air-Microwave Drying of Foods," *Proceedings of the 7th International Conference on Microwave and High Frequency Heating*, Valencia, Spain, 1999, pp. 43–48.

[23] Funebo, T., and T. Ohlsson, "Dielectric Properties of Fruits and Vegetables as a Function of Temperature and Moisture Content," *Int. J. Microwave Power and Electromagnetic Energy*, Vol. 34, No. 1, 1999, pp. 42–54.

[24] Bengtsson, N. E., and P. O. Risman, "Dielectric Properties of Foods at 3 GHz as Determined by a Cavity Perturbation Technique," *Int. J. Microwave Power and Electromagnetic Energy*, Vol. 6, No. 2. 1971, pp. 107–123.

[25] Zhang, Q., T. H. Jackson, and A. Ungan, "Numerical Modeling of Microwave Induced Natural Convection," *Int. J. Heat and Mass Transfer*, Vol. 43, 2000, pp. 2141–2154.

[26] Heldmann, R., (ed.), *Encyclopedia of Agricultural, Food and Biological Engineering*, New York: Marcel Dekker, 2003.

[27] Wang, S., M. Monzon, and Y. Gazit, et al., "Temperature Dependent Dielectric Properties of Selected Subtropical and Tropical Fruits and Associated Insect Pests," *Transactions on ASAE*, Vol. 48, No. 5, 2005, pp. 1873–1881.

[28] Wang, Y., J. Tang, and B. Rasco, et al., "Dielectric Properties of Salmon Fillets as a Function of Temperature and Composition," *Journal of Food Engineering*, Vol. 87, 2008, pp. 236–246.

[29] Guo, W., G. Tiwari, and J. Tang, et al., "Frequency, Moisture and Temperature Dependent Dielectric Properties of Chickpea Flour," *Biosystems Engineering*, Vol. 101, 2008, pp. 217–224

[30] Liu, Y., J. Tang, and Z. Mao, "Analysis of Bread Dielectric Properties Using Mixture Equations," *Journal of Food Engineering*, Vol. 93, 2009, pp. 72–79.

[31] Wang, J., J. Tang, and Y. Wang, et al., "Dielectric Properties of Egg Whites and Whole Egg as Influenced by Thermal Treatments," *LWT—Food Science and Technology*, Vol. 42, 2009, pp. 1204–1212.

4

Biological Tissues

The main mechanism for the interaction between EM waves and biological tissues is the same as in foodstuff: oscillation of polar water molecules (H_2O) and ions. Part of the water in biotissues is linked with albumens (0.3~0.4g of water on 1g of albumen) and at relaxation frequencies specific electrical conductivity (σ) of bound water is sometimes higher than σ of pure water. The loss factor of free (ε''_{fw}) and bound (ε''_{bw}) water in biological tissues is expressed as [1]:

$$\varepsilon''_{fw} \approx 20\varepsilon''_{bw}, \text{ at 500 MHz; } \varepsilon''_{fw} \approx 0.1\varepsilon''_{bw}, \text{ at 2450 MHz}$$

4.1 Human Body Tissues

Microwave energy is widely used in medical treatments including physical therapy, diagnostics, rapid rewarming of cryopreserved tissues, pharmacology, reflex therapy, blood sterilization, and hyperthermal treatment of cancer [2–5].

Human biological tissues are divided in two main groups: materials with high and low water contents. Muscles (73%~78%), liver (75%~77%), kidneys (76%~78%), brain (68%~73%), skin

(60%~76%), lung (80%~83%), and eye (up to 89%) compile the first group. Fat (5%~10%) and bone (8%~16%) tissues may be included in the second group. Moisture content in blood is higher than in other tissues (up to 83%). Dielectric properties of human blood at different temperatures are represented in Table 4.1. Blood flow usually does not influence their dielectric properties, excluding tissues with high blood volume (kidneys) or low moisture content (fat).

Table 4.1
Complex Dielectric Permittivity and Conductivity
of Human Blood at 2.45 GHz

$T°C$	Dielectric Permittivity	Loss Factor	Electrical Conductivity (S/m)
15	59.9	19.9	2.71
25	57.5	17.1	2.33
35	56	15.9	2.166

From: [6].

Dielectric permittivity and electrical conductivity of some biological materials at room temperature and ISM frequencies taken from [7] are represented in Table 4.2. It is interesting to compare this data with the CDP values of human tissues obtained in [6, 8, 9] at different microwave frequencies (Tables 4.3–4.5).

Investigations of cancerous biological samples show [10] that both ε' and ε'' of a tumor is higher than the normal tissue because of higher water content (Table 4.6).

Table 4.2
Dielectric Permittivity and Electrical Conductivity (σ, S/m) of Human Tissues at Room Temperature

Tissue	433 MHz ε'	433 MHz σ	915 MHz ε'	915 MHz σ	2,450 MHz ε'	2,450 MHz σ	5,800 MHz ε'	5,800 MHz σ
Bladder	19.622	0.329	18.922	0.385	18.00	0.685	16.243	1.857
Blood	63.837	1.361	61.314	1.544	58.264	2.545	52.539	6.505
Bone, cancellous	22.262	0.241	20.756	0.344	18.549	0.805	15.395	2.148
Bone, cortical	13.074	0.094	12.439	0.145	11.381	0.394	9.675	1.154
Bone marrow, infiltrated	11.802	0.185	11.261	0.229	10.308	0.459	8.713	1.149
Breast fat	5.507	0.035	5.422	0.049	5.147	0.137	4.497	0.419
Cartilage	45.149	0.598	42.6	0.789	38.711	1.756	32.15	4.89
Cerebellum	55.132	1.047	49.348	1.269	44.804	2.101	39.98	4.995
Cerebrospinal fluid	70.639	2.259	68.606	2.419	66.243	3.458	60.469	7.839
Colon (large intestine)	62.015	0.873	57.866	1.087	53.878	2.038	48.455	5.569
Cornea	58.775	1.206	55.171	1.40	51.614	2.295	46.532	5.664
Dura	46.376	0.835	44.391	0.965	42.035	1.669	37.876	4.309
Eye tissue (sclera)	57.380	1.014	55.23	1.172	52.627	2.033	47.806	5.465
Fat	5.567	0.042	5.459	0.051	5.280	0.1045	4.955	0.293
Gall bladder	60.892	1.144	59.118	1.261	57.634	2.059	53.679	5.723
Grey matter	56.829	0.751	52.653	0.948	48.911	1.807	44.004	4.986
Heart	65.329	0.983	59.796	1.238	54.814	2.256	48.949	5.862
Kidney	65.454	1.116	58.557	1.40	52.743	2.429	46.753	5.896
Lens cortex	47.961	0.675	46.545	0.798	44.625	1.504	40.686	4.369
Lens nucleus	37.293	0.379	35.814	0.489	33.973	1.087	30.471	3.428
Liver	50.689	0.667	46.764	0.861	43.034	1.686	38.13	4.642
Lung, inflated	23.585	0.379	21.972	0.459	20.477	0.804	18.508	2.077
Lung, deflated	54.197	0.694	51.372	0.864	48.381	1.682	43.75	4.821
Muscle (parallel fiber)	58.588	0.847	56.845	1.001	54.417	1.882	49.498	5.44
Muscle (transverse fiber)	56.873	0.805	54.997	0.948	52.729	1.739	48.484	4.962
Nerve	35.048	0.455	32.485	0.577	30.145	1.088	27.217	2.941
Ovary	56.688	1.032	50.385	1.298	44.699	2.264	38.691	5.284
Skin, dry	46.079	0.702	41.329	0.872	38.007	1.464	35.114	3.717
Skin, wet	49.418	0.681	46.021	0.850	42.852	1.592	38.624	4.342
Spleen	62.453	1.043	57.087	1.280	52.449	2.238	46.942	5.672
Stomach	67.194	1.013	65.019	1.193	62.158	2.211	56.475	6.314
Tendon	47.125	0.568	45.796	0.724	43.121	1.684	36.751	5.254
Testis prostate	63.031	1.037	60.506	1.215	57.551	2.167	52.213	5.948
Thyroid thymus	61.331	0.886	59.649	1.044	57.200	1.967	52.052	5.721
Tongue	57.380	0.783	55.23	0.942	52.627	1.803	47.806	5.235
Trachea	43.935	0.644	41.971	0.776	39.733	1.448	35.943	4.079
Uretus	64.048	1.087	61.061	1.276	57.814	2.246	52.353	6.047
Vitreous humor	68.995	1.534	68.898	1.641	68.208	2.478	64.778	6.674
White matter	41.671	0.454	38.837	0.595	36.167	1.215	32.621	3.494

From: [7].

Table 4.3

Dielectric Properties of Human Body Tissues at 37°C
and Microwave Frequencies

Tissue	Comment	Frequency (GHz)	ε'	ε''
Brain	In vitro	2.45	32	15.5
		3	33	18
		4	33	15.8
Marrow	In vitro	3	4.5 ÷ 5.8	0.7 ÷ 1.35
Eye	Lens	2.45	30	8
		3	30	9
		5	30	10
Fat	In vitro	2.45	5.75	0.8
		3	3.9 ÷ 7.2	0.67 ÷ 1.36
		5	4.7	0.7
	Breast	3	3.94	0.8
	Fistula	3	7	1.75
Liver	In vitro	3	42	12.2
Muscle	In vitro	1	49 ÷ 52	23.5
		2.45	47.5	13.5
		3	45 ÷ 48	13÷14
		5	44	14
Skin	In vitro	1	43 ÷ 46	16.4 ÷ 20
		2.45	43	14
		3	40 ÷ 45	12 ÷ 16

From: [6].

Table 4.4
Dielectric Properties of Some Human Tissues at $T = 37°C$

Tissue	433 MHz ε'	σ(S/m)	915 MHz ε'	σ(S/m)	2,450 MHz ε'	σ(S/m)
Blood	66	1.27	62	1.41	60	2.04
Bone with marrow	5.2	0.11	4.9	0.15	4.8	0.21
Brain (white matter)	48	0.63	41	0.77	35.5	1.04
Brain (grey matter)	57	0.83	50	1.0	43	1.43
Fat	15	0.26	15	0.35	12	0.82
Kidney	60	1.22	55	1.41	50	2.63
Liver	47	0.89	46	1.06	44	1.79
Muscle	57	1.12	55.4	1.45	49.6	2.56
Skin	47	0.84	45	0.97	44	1.85
Ocular tissue (choroid)	60	1.32	55	1.4	52	2.3
Ocular tissue (cornea)	55	1.73	51.5	1.9	49	2.5
Ocular tissue (iris)	59	1.18	55	1.18	52	2.1
Ocular tissue (lens cortex)	55	0.8	52	0.97	48	1.75
Ocular tissue (lens nucleus)	31.5	0.29	30.8	0.5	26	1.4
Ocular tissue (retina)	61	1.5	57	1.55	56	2.5

From: [8].

Table 4.5
Dielectric Properties of Body Tissues at 915 MHz

Tissue	Conductivity (S/m)	Relative Permittivity	Loss Factor
Aorta	0.7009	44.741	13.7686
Bladder	0.38506	18.923	7.5646
Blood	1.5445	61.314	30.342
Bone, cancellous	0.3435	20.756	6.7488
Bone, cortical	0.14512	12.44	2.851
Bone marrow	0.0406	5.5014	0.797
Brain, grey matter	0.94866	52.654	18.636
Brain, white matter	0.5954	38.873	11.708
Breast fat	0.049523	5.4219	0.9729
Cartilage	0.7892	42.6	15.504

Table 4.5 (continued)

Tissue	Conductivity (S/m)	Relative Permittivity	Loss Factor
Cerebellum	0.0025	49.349	0.5054
Cerebrospinal fluid	2.4187	68.607	47.517
Cervix	0.96402	49.748	18.9386
Colon	1.087	57.867	21.353
Cornea	1.4	55.172	27.5186
Duodenum	1.1932	65.02	23.4416
Dura	0.96584	44.391	18.9745
Eye sclera	1.1725	55.23	23.034
Fat	0.051402	5.4596	1.009
Gall bladder	1.2614	59.118	24.781
Gland	1.0444	59.65	20.517
Heart	1.2378	59.796	24.3178
Kidney	1.4007	58.556	27.516
Lens	0.7979	46.545	15.675
Liver	0.86121	46.764	16.9187
Lung,deflated	0.86373	51.373	16.968
Lung,inflated	0.45926	21.972	9.022
Mucous membrane	0.85015	46.021	16.701
Muscle	0.94809	54.997	18.625
Nail	0.14512	12.44	2.851
Nerve	0.57759	32.486	11.347
Ovary	1.2985	50.36	25.509
Prostate	1.2159	60.506	23.8859
Retina	1.1725	55.23	23.0342
Skin, dry	0.8717	41.329	17.087
Skin, wet	0.85015	46.021	16.7015
Spleen	1.2801	57.087	25.14796
Stomach	1.1932	65.02	23.441
Tendon	0.72441	45.796	14.231
Testis	1.2159	60.506	23.8859
Tongue	0.94198	55.23	18.505
Tooth	0.14512	12.44	2.851
Trachea	0.7757	41.971	15.238
Uterus	1.2764	61.061	25.076

From: [9].

The thermal field distribution in microwave-exposed biological tissue is defined by Pennes's bioheat equation [11]:

$$\rho_t C_t \frac{\partial T}{\partial \tau} = \lambda_t \nabla^2 T - \rho_t \rho_b C_b F \left(T - T_b \right) + Q_m + Q_v \tag{4.1}$$

where ρ_t, ρ_b are the densities of the tissue and the blood correspondently, C_t, C_b are the heat capacities of the tissue and the blood, λ_t is the thermal conductivity of the tissue, F is the blood flow rate, Q_m is the specific power density caused by biochemical processes inside human body, and T_b is the blood temperature. The thermal parameters of some body tissues are given in Table 4.7.

Table 4.6
Dielectric Properties of Human Breast Tissues at Room Temperature
and Frequency 2,983 MHz

Sample	Comment	ε'	σ(S/m)	Bound Water Content (%)
Breast tissue, Patient 1 (age 47)	Normal	20.43	3.12	43
	Cancerous	32.31	3.52	62
Breast tissue, Patient 2 (age 49)	Normal	18.85	2.71	42
	Cancerous	38.73	4.12	65
Breast tissue, Patient 3 (age 51)	Normal	24.98	3.25	45
	Cancerous	40.1	4.31	65
Breast tissue, Patient 4 (age 45)	Normal	19.5	2.64	41
	Cancerous	30	3.34	61

From: [10].

Table 4.7
Thermophysical Constants of Some Human Tissues at Room Temperature

Parameters	Blood	Muscle	Liver	Skin	Fat	Brain
Density (kg/m^3)	1,060	1,020	1,070	1,100	916	1,030
Heat capacity (J/(kg·K))	3,960	3,500	3,590	3,500	2,300	3,640
Thermal conductivity (W/(m·K))	0.61	0.6	0.488	0.5	0.22	0.53

From: [11].

4.2 Animals and Insects

Intensive studies of the effects of microwave radiation on different biological tissues are necessary for preventing negative consequences of microwaves influence on the human body. In particular, ocular effects due to microwave energy have been investigated in [12, 13]. Information about the CDP and thermal properties of animal tissues is very important both in experimental and theoretical studies, most of which are carried out at room temperature (Tables 4.8 and 4.9).

Table 4.8
Thermal Properties of Rabbit Tissues at Room Temperature

Parameter	Units	Cornea and Sclera	Anterior Chamber	Lens	Vitreous Humor
Heat capacity	J/kg·K	4,178	3,997	3,000	3,997
Thermal conductivity	W/m·K	0.58	0.603	0.4	0.603
Density	kg/m³	1,050	1,000	1,050	1,000

From: [13].

Table 4.9
Dielectric Properties of Normal and Polluted Wistar Rat Tissues at Room Temperature and 2.45 GHz

Rat Tissue	Normal Tissue ε'	σ(S/m)	Tissue Treated with Lead ε'	σ(S/m)	Tissue Treated with Cadmium ε'	σ(S/m)
Liver	42.6 ± 0.96	1.52 ± 0.08	45.4 ± 1.38	1.2 ± 0.16	42.5 ± 0.8	1.35±0.18
Lung	47.36 ± 0.69	1.64 ± 0.09	38.9 ± 1.34	1.19 ±0.1	40.8 ± 1.3	1.19±0.09
Kidney	49.84 ± 1.09	1.77 ± 0.21	45.7 ± 1.18	1.21 ±0.2	44.2 ± 1.0	1.08 ± 0.1
Pancreas	41.7 ± 0.78	1.58 ± 0.09	46.8 ± 1.23	1.4 ± 0.14	51.2 ± 1.0	1.2 ± 0.13
Muscle	49.5 ± 0.64	1.77 ± 0.05	48.1 ± 0.69	1.19 ± 0.2	45.7 ± 1.4	1.1 ± 0.09

From: [15].

Results of measurements obtained in [14] have shown that ε' and ε'' of biotissues in vivo (directly in organism) and in vitro (in test tube) almost coincide at frequencies higher than 100 MHz. Both human and animal tissue is usually heated by microwaves only up to 39 ÷ 41°C because at higher temperatures it can be damaged. The CDPs of some animal tissue at different temperatures are given in Table 4.10.

Table 4.10
Dielectric Properties of Animal Tissues at Microwave Frequencies

Animal	Tissue	Comment	$T°C$	f (GHz)	ε'	ε''
Dog	Grey matter	In vivo	25	2	50	11
			37		49	12
	White matter	In vivo	25	2	37	8.5
			37		37	8.7
	Fat	In vitro	37	2.45	12	5.1
	Kidney	In vivo	37	1	49.5	14.4
				2.45	47.5	13.2
				3	47	12
				5	43	14
	Muscle	In vivo	34	1	47	20
				2.45	45	11
Rat	Muscle	In vivo	31	1	61	23.5
				2.45	58	17.5
				3	56	17.3
				5	53	19.2
	Brain	In vivo	32	1	55	21.5
				2.45	52.5	14
				3	52.5	13.7
				5	53	16
	Blood	In vivo	23	1	62	28
				2.45	62	18
				3	62	19

From: [6].

The application of EM waves for rapid rewarming of cryopreserved tissues offers some advantages over conventional methods in medical technology such as organ transplantation. Information about the CDP of rabbit and dog kidney tissues at low and room temperature (Tables 4.11 and 4.12) can be useful in modeling and designing microwave heating systems intended for such purposes. Here we observe the same effect as for food materials: low values of dielectric permittivity at low temperatures and rapid increasing of this parameter at room temperature.

Table 4.11
Dielectric Properties of Rabbit Kidney Tissues
at 2,450 MHz

Temperature (°K)	Dielectric Permittivity	Loss Factor	Conductivity (mS/cm)
239	2.2	0.506	0.7
253	4.3	1.29	1.7
273	49	14.21	20
293	56	15.68	28
295	57	17.1	30

From: [15].

Table 4.12
Dielectric Properties of Dog Kidney Tissues
at 918 MHz

Temperature (°K)	Dielectric Permittivity	Loss Factor	Conductivity (mS/cm)
240	7	0.78	0.04
253	9	1.37	0.07
263	16	4.31	0.22
272	54	12.53	0.64
278	56	16.06	0.82
285	55	17.8	0.91
290	54	18.6	0.95
298	53	19.2	0.98

From: [16].

Microwave energy has been successfully utilized for the destruction of insects in wood samples [17] and postharvest crops [18]. Temperatures up to 53°C are enough for this purpose. Data on the CDP of some insect larvae is represented in Table 4.13 for temperatures $20 \leq T°C \leq 60$. As one can see from this data, both the dielectric permittivity and loss factor of all considered insects remain almost constant at 915 and 1,800 MHz. Additional information about the dielectric properties of grain weevils and potato beetles can be found in [19, 20].

Table 4.13
Dielectric Properties (Mean ± STD of Two Replicates) of Four Insect Larvae
at Two Frequencies

Biomaterial	f (MHz)	$T°$C	20	30	40	50	60
Colding moth	915	ε'	47.9±0.2	45.9±0.9	44.6±0.6	45.6±1.5	45±2.4
		ε''	11.7±0.1	12.5±0.4	13.9±0.5	16.5±0.3	19.1±1
	1,800	ε'	44.5±0.1	42.9±0.9	41.6±0.4	42.7±1.5	41.9±2.2
		ε''	12±0.2	11.7±0.3	11.9±0.3	13.2±0.3	14.2±0.7
Indian meal moth	915	ε'	39.9±0.4	39.2±0.0	37.6±0.8	37.2±1.3	37.8±1.6
		ε''	13.4±1.4	14.3±1.4	15.2±2.1	16.9±2.4	11.4±1.5
	1,800	ε'	37.5±0.5	36.9±0.1	35.5±0.9	35.3±1.4	35.6±1.7
		ε''	10.6±0.6	10.6±0.8	10.6±1.2	11.4±1.5	12.8±1.7
Mexican fruit fly	915	ε'	48.5±3.4	47.3±3.5	46.4±2.9	45.7±2.3	44.5±2.0
		ε''	17.5±2.0	21.3±3.9	24.2±5.1	26.8±5.7	15.4±2.5
	1,800	ε'	47.0±0.7	45.4±0.4	44.7±0.8	44.1±1.4	43.0±1.6
		ε''	13.3±1.7	13.9±1.9	14.5±2.2	15.4±2.5	16.5±2.7
Navel orange worm	915	ε'	44.5±1.3	43.6±0.4	42.8±0.1	42.3±0.4	42.2±0.1
		ε''	16.1±0.1	17.5±0.6	19.2±0.9	21.2±1.0	24±0.1
	1,800	ε'	42.2±1.4	41.5±0.7	40.7±0.1	40.2±0.1	40.0±0.0
		ε''	12.7±0.0	12.9±0.4	13.4±0.6	14.1±0.6	15.5±0.0

From: [18].

References

[1] Adamenko, V. Y., and V. K. Mamishkni, "Bound Water in Biologically Active Frequency Range," *Modern Problems of Microwave Energy Application*, Proceedings of the Conference, Saratov, Russia, 1993, pp. 47–48 (in Russian).

[2] Lu, C. C., H. Z. Li, and D. Gao, "Combined Electromagnetic and Heat Conduction Analysis of Rapid Rewarming of Cryopreserved Tissues," *IEEE Transactions on Microwave Theory and Techniques*, Vol. MTT-48, No. 11, 2000, pp. 2185–2190.

[3] Gibbs, F. A., "Clinical Evaluation of a Microwave/Radiofrequency System (BSD Corporation) for Induction of Local and Regional Hyperthermia," *Journal of Microwave Power and Electromagnetic Energy*, Vol. 16, No. 2, 1981, pp. 185–192.

[4] Guy, A. W., J. F. Lehmann, and J. B. Stonebridge, "Therapeutic Applications of Electromagnetic Power," *IEEE Proceedings*, Vol. 62, No. 1, 1974, pp. 55–75.

[5] Iskander, M. F., and C. H. Durney, "Microwave Methods of Measuring Changes in Lung Water," *Journal of Microwave Power and Electromagnetic Energy*, Vol. 18, No. 3, 1983, pp. 265–275.

[6]   Stuchly, M. A., and S. S. Stuchly, "Dielectric Properties of Biological Substances—Tabulated," *Journal of Microwave Power and Electromagnetic Energy*, Vol. 15, No.1, 1980, pp. 19–26.

[7]   Gabriel, C., *Compilation of the Dielectric Properties of Body Tissues at RF and Microwave Frequencies*, Brooks Air Force Technical Report, AL/OE-TR-1996-0037.

[8]   Pethig, R., "Dielectric Properties of Biological Materials: Biophysical and Medical Applications," *IEEE Transactions on Electrical Insulation*, Vol. EI-19, No. 5, 1984, pp. 453–474.

[9]   Italian National Research Council, Institute for Applied Physics .

[10] Bingu, G., S. J. Abraham, and A. Lonappan, et al., "Detection of Dielectric Contrast of Breast Tissues Using Confocal Microwave Technique," *Microwave and Optical Technology Letters*, Vol. 48, No. 6, 2006, pp. 1187–1190.

[11] Kikuchi, S., K. Saito, and M. Takahashi, et al., "Control of Heating Pattern for Interstitial Microwave Hyperthermia by a Coaxial-Dipole Antenna—Aiming at Treatment of Brain Tumor," *Electronics and Communications in Japan*, Vol. 90, No. 12, 2007, pp. 31–38.

[12] Birenbaum, L., I. T. Kaplan, and W. Metlay, et al., "Effects of Microwaves on the Rabbit Eye," *Journal of Microwave Power and Electromagnetic Energy*, Vol. 4, N. 4, 1969, pp. 232-243.

[13] Wake, K., H. Hongo, and S. Watanabe, et al., "Development of a 2.45 GHz Local Exposure System for In Vivo Study on Ocular Effect," *IEEE Transactions on Microwave Theory and Techniques*, Vol. MTT-55, No. 3, 2007, pp. 588–596.

[14] Kraszewski, A., M. A. Stuchly, and S. S. Stuchly, et al., "In Vivo and In Vitro Dielectric Properties of Animal Tissues at Radio Frequencies," *Electromagnetics*, Vol. 3, 1982, pp. 421–432.

[15] Sebastian, J. L., S. Munoz, and J. M. Miranda, et al., "A Simple Experimental Set-Up for the Determination of the Complex Dielectric Permittivity of Biological Tissues at Microwave Frequencies," *Proceedings of the 34th European Microwave Conference*, Amsterdam, Netherlands, 2004, pp. 661–663.

[16] Burgette, E. C., and A. M., Karow, "Kidney Model for Study Electromagnetic Thawing," *Cryobiology*, Vol. 15, No. 2, 1978, pp. 142–151.

[17] Andreuccetti, D., M. Bini, and A. Ignesti, et al., "Microwave Destruction of Woodworms," *Journal of Microwave Power and Electromagnetic Energy*, Vol. 29, No. 3, 1994, pp. 153–160.

[18] Wang, S., J. Tang, and J. A. Johnson, et al., "Dielectric Properties of Fruits and Insect Pests as Related to Radio Frequency and Microwave Treatments," *Biosystems Engineering*, Vol.85, No. 2, 2003, pp. 201–212.

[19] Nelson, S. O., and J. A. Payne, "RF Dielectric Heating for Pecan Weevil Control," *Transactions of the ASABE*, Vol. 25, No. 8, 1982, pp. 456–458.

[20] Colpitts, B., Y. Pelletier, and S. Cogswell, "Complex Permittivity Measurements of the Colorado Potato Beetle Using Coaxial Probe Techniques," *Journal of Microwave Power and Electromagnetic Energy*, Vol. 27, No. 3, 1992, pp. 175–182.

5

Fibrous Materials

Most fibrous materials are organic cellulose based substances such as wood, paper, carton, fabric, and fibers. Wood is highly hygroscopic and anisotropic. The dielectric properties of wood depend on the type of wood, density, moisture content, and temperature. The normal moisture content of wood is between 8% and 12%.

According to the generalized electrophysical model of wood samples proposed in [1], one can consider a second-order tensor for evaluation of its complex dielectric permittivity:

$$\left\|\varepsilon' - j\varepsilon''\right\| = \begin{vmatrix} \varepsilon'_{LL} - j\varepsilon''_{LL} & \varepsilon'_{LR} - j\varepsilon''_{LR} & \varepsilon'_{LU} - j\varepsilon''_{LU} \\ \varepsilon'_{RL} - j\varepsilon''_{RL} & \varepsilon'_{RR} - j\varepsilon''_{RR} & \varepsilon'_{RU} - j\varepsilon''_{RU} \\ \varepsilon'_{UL} - j\varepsilon''_{UL} & \varepsilon'_{UR} - j\varepsilon''_{UR} & \varepsilon'_{UU} - j\varepsilon''_{UU} \end{vmatrix} \tag{5.1}$$

where L, R, and U are the longitudinal, radial, and tangential axes of anisotropy, respectively. Rotation of electric field vector $(\bar{E})$ on 180° does not change the dielectric properties of wood materials. That is, L, R, and U are the principle axes of anisotropy, and the tensor in (5.1) may be simplified as:

$$\|\varepsilon' - j\varepsilon''\| = \begin{vmatrix} \varepsilon'_L - j\varepsilon''_L & 0 & 0 \\ 0 & \varepsilon'_R - j\varepsilon''_R & 0 \\ 0 & 0 & \varepsilon'_U - j\varepsilon''_U \end{vmatrix} \qquad (5.2)$$

When $\bar{E}$ is arbitrarily oriented in space and forms angle ϑ_1 with L, angle ϑ_2 with R, and angle ϑ_3 with U, closed-form expressions for calculation of ε' and $\tan \delta_e$ are derived in [1]:

$$\varepsilon' = \varepsilon'_L \cos^2 \vartheta_1 + \varepsilon'_R \cos^2 \vartheta_2 + \varepsilon'_U \cos^2 \vartheta_3 \qquad (5.3)$$

$$\tan \delta_e = tg\delta_{eL} \cos^2 \vartheta_1 + tg\delta_{eR} \cos^2 \vartheta_2 + tg\delta_{eU} \cos^2 \vartheta_3 \qquad (5.4)$$

The operating frequency 2.45 GHz is widely utilized in many industrial microwave applicators. That is why information about the CDP of wood at this frequency as a function of temperature, moisture, and pressure (p) is usually available in the literature [2–4] for different types of wood (see Tables 5.1–5.3). The data in Table 5.2 shows the parallel orientation of wood fibers and electric field vector.

Table 5.1
Dielectric Properties of Mountain Ash (Eucalyptus regnans) at 2.45 GHz

Temperature, °C	Expression	R^2
$20 \leq T \leq 100$ Normal pressure	$\varepsilon'(T) = -4 \cdot 10^{-6}T^3 + 0.0002T^2 - 0.0212T + 21.483$	0.9788
	$\varepsilon''(T) = 3 \cdot 10^{-7}T^3 - 0.0003T^2 - 0.004T + 7.3238$	0.9653
Pressure, bars	**Expression**	R^2
$0 \leq p \leq 5.5$ Room temperature	$\varepsilon'(p) = -0.0585p^3 + 0.3181p^2 - 2.2826p + 20.241$	0.9836
	$\varepsilon''(p) = -0.1102p^3 + 1.1708p^2 - 4.1352p + 5.9146$	0.9964

From: [2].

Table 5.2
Dielectric Properties of Douglas Fir at 2,450 MHz
in Temperature Range $20 \leq T°C \leq 90$

W%	Dielectric Permittivity and Loss Factor	R^2
0	$\varepsilon'(T) = -5 \cdot 10^{-7}T^3 + 8 \cdot 10^{-5}T^2 - 0.0027T + 2.058$	0.9156
	$\varepsilon''(T) = 0.0013T - 0.0219$	0.9595
10	$\varepsilon'(T) = -2 \cdot 10^{-6}T^3 + 0.0004T^2 + 0.0007T + 2.1389$	0.9978
	$\varepsilon''(T) = -8 \cdot 10^{-5}T^2 + 0.0103T + 0.1857$	0.9768
20	$\varepsilon'(T) = 3 \cdot 10^{-7}T^3 - 0.0002T^2 + 0.047T + 2.6054$	0.9982
	$\varepsilon''(T) = -9 \cdot 10^{-5}T^2 + 0.0051T + 0.8268$	0.9961
30	$\varepsilon'(T) = 3 \cdot 10^{-6}T^3 - 0.0007T^2 + 0.0878T + 3.5821$	0.9991
	$\varepsilon''(T) = -3 \cdot 10^{-5}T^2 - 0.007T + 1.4894$	0.9933

From: [3].

Table 5.3
Dielectric Properties of Wood Samples at 2.45 GHz and
Room Temperature $0 \leq W\% \leq 25$

Sample	Expression	R^2
Sylvester pine	$\varepsilon'(W) = 9 \cdot 10^{-5}W^3 - 0.0008W^2 + 0.0408W + 1.7861$	0.9996
	$\varepsilon''(W) = -5 \cdot 10^{-5}W^3 + 0.0027W^2 - 0.0069W + 0.0258$	0.9971
Poplar	$\varepsilon'(W) = 0.0004W^2 + 0.0485W + 1.6526$	0.9993
	$\varepsilon''(W) = -2 \cdot 10^{-5}W^3 + 0.0011W^2 + 0.0095W + 0.0217$	0.9990
Chestnut	$\varepsilon'(W) = 2 \cdot 10^{-5}W^3 + 0.0013W^2 + 0.0484W + 1.8851$	0.9998
	$\varepsilon''(W) = -10^{-4}W^3 + 0.0044W^2 - 0.0128W + 0.0296$	0.9946
Oak	$\varepsilon'(W) = 10^{-4}W^3 - 0.0004W^2 + 0.0644W + 2.0473$	0.9999
	$\varepsilon''(W) = -10^{-4}W^3 + 0.005W^2 - 0.0164W + 0.0527$	0.9984
Walnut	$\varepsilon'(W) = 2 \cdot 10^{-5}W^3 + 0.0015W^2 + 0.0403W + 1.8689$	0.9994
	$\varepsilon''(W) = -2 \cdot 10^{-5}W^3 + 0.0022W^2 - 0.0003W + 0.0253$	0.9991

From: [4].

As it has been shown in [1] CDP of wood depends not only on temperature and moisture but also wood structure and density. Data in Tables 5.4–5.6 for three frequencies are represented for the case when electric field vector orientation is perpendicular to the grain.

Table 5.4

Dielectric Properties of Dry and Moist Wood Samples at 1 GHz

| W% | T°C | Density in Oven-Dry Conditions (g/cm³) | | | | | | | | | | | |
| | | 0.3 | | 0.4 | | 0.5 | | 0.6 | | 0.7 | | 0.8 | |
		ε'	ε''	ε'	ε''	ε'	ε''	ε'	ε''	ε'	ε''	ε'	ε''
0	20	1.5	0.025	1.7	0.037	1.8	0.045	2.0	0.058	2.2	0.07	2.3	0.059
	50	1.6	0.030	1.8	0.045	1.9	0.053	2.1	0.067	2.3	0.082	2.4	0.096
	90	1.7	0.037	1.9	0.053	2.0	0.064	2.2	0.081	2.4	0.098	2.5	0.115
10	20	1.9	0.114	2.2	0.176	2.5	0.25	2.7	0.324	3.0	0.42	3.3	0.528
	50	2.0	0.12	2.3	0.184	2.7	0.27	2.9	0.348	3.2	0.448	3.5	0.56
	90	2.2	0.132	2.5	0.2	2.9	0.29	3.1	0.372	3.4	0.476	3.8	0.608
20	20	2.4	0.216	2.9	0.348	3.4	0.544	3.9	0.741	4.4	0.924	4.8	1.2
	50	2.44	0.195	2.9	0.319	3.45	0.518	4.0	0.72	4.5	0.9	4.9	1.176
	90	2.5	0.20	3.0	0.3	3.5	0.49	4.0	0.64	4.55	0.819	5.0	1.1
30	20	3.2	0.416	3.9	0.663	4.7	1.034	5.4	1.404	6.1	1.83	6.8	2.312
	50	3.2	0.384	3.9	0.624	4.7	0.94	5.4	1.296	6.1	1.708	6.8	2.108
	90	3.2	0.352	3.9	0.546	4.7	0.846	5.4	1.134	6.1	1.525	6.8	1.708

From: [1].

Table 5.5

Dielectric Properties of Dry and Moist Wood Samples at 2.4 GHz

| W% | T°C | Density in Oven-Dry Conditions (g/cm³) | | | | | | | | | | | |
| | | 0.3 | | 0.4 | | 0.5 | | 0.6 | | 0.7 | | 0.8 | |
		ε'	ε''	ε'	ε''	ε'	ε''	ε'	ε''	ε'	ε''	ε'	ε''
0	20	1.4	0.021	1.6	0.03	1.7	0.039	1.9	0.051	2.1	0.063	2.2	0.073
	50	1.5	0.026	1.7	0.036	1.8	0.047	2.0	0.06	2.2	0.075	2.3	0.085
	90	1.55	0.029	1.8	0.043	1.9	0.055	2.1	0.074	2.3	0.087	2.4	0.1
10	20	1.8	0.108	2.0	0.16	2.3	0.253	2.5	0.325	2.8	0.42	3.0	0.51
	50	1.9	0.114	2.1	0.168	2.4	0.264	2.7	0.351	3.0	0.45	3.2	0.544
	90	2.05	0.123	2.3	0.184	2.6	0.286	2.85	0.371	3.2	0.48	3.4	0.578
20	20	2.1	0.21	2.5	0.35	2.9	0.522	3.3	0.693	3.7	0.925	4.0	1.12
	50	2.1	0.189	2.5	0.325	2.9	0.493	3.4	0.646	3.8	0.874	4.1	1.025
	90	2.2	0.176	2.6	0.286	3.0	0.45	3.4	0.578	3.8	0.76	4.1	0.943
30	20	2.7	0.351	3.2	0.576	3.8	0.836	4.3	1.161	4.9	1.519	5.4	1.89
	50	2.7	0.324	3.2	0.512	3.8	0.76	4.3	1.032	4.9	1.372	5.4	1.728
	90	2.7	0.27	3.2	0.448	3.8	0.646	4.3	0.903	4.9	1.176	5.4	1.458

From: [1].

Table 5.6
Dielectric Properties of Dry and Moist Wood Samples at 5.8 GHz

		Density in Oven-Dry Condition (g/cm^3)										
		0.3		0.4		0.5		0.6		0.7		0.8	
W%	T°C	ε'	ε''	ε'	ε''	ε'	ε''	ε'	ε''	ε'	ε''	ε'	ε''
0	20	1.4	0.019	1.5	0.027	1.7	0.037	1.8	0.047	2.0	0.058	2.2	0.07
	50	1.4	0.022	1.55	0.033	1.75	0.044	1.85	0.056	2.1	0.069	2.3	0.085
	90	1.5	0.029	1.6	0.038	1.8	0.054	1.9	0.067	2.1	0.082	2.35	0.101
10	20	1.7	0.102	1.9	0.143	2.15	0.215	2.3	0.276	2.5	0.35	2.7	0.432
	50	1.8	0.108	2.0	0.16	2.3	0.23	2.4	0.288	2.7	0.378	2.9	0.464
	90	1.9	0.114	2.2	0.176	2.5	0.25	2.6	0.312	2.9	0.406	3.1	0.496
20	20	2.0	0.24	2.3	0.368	2.6	0.546	3.0	0.75	3.4	0.986	3.6	1.188
	50	2.1	0.231	2.4	0.36	2.7	0.513	3.1	0.713	3.5	0.945	3.7	1.147
	90	2.1	0.21	2.5	0.325	2.8	0.476	3.2	0.64	3.6	0.864	3.85	1.04
30	20	2.7	0.432	3.3	0.693	3.8	0.988	4.4	1.408	4.9	1.813	5.5	2.365
	50	2.8	0.42	3.4	0.646	3.9	0.936	4.5	1.305	5.1	1.734	5.7	2.223
	90	2.9	0.377	3.5	0.56	4.1	0.820	4.7	1.175	5.2	1.508	5.9	2.0

From: [1].

Sometimes information about the CDP of wood at microwave frequencies close to ISM bands can be also useful. Table 5.7 shows dependencies $\varepsilon'(M)$ and $\varepsilon''(M)$ of four different wood samples at 1,110 MHz, where M is the moisture content defined as:

$$M = (M_m - M_d)/V \qquad (5.5)$$

here M_m is the moist wood mass, M_d is the dry wood mass, and V is the sample volume.

Another example is the temperature-dependent parameters of pine sapwood at 922 MHz (see Table 5.8).

Table 5.7

Dielectric Properties of Some Wood Samples at 1,110 MHz and Temperature 24°C

Sample	Moisture	Dielectric Permittivity and Loss Factor	R^2
Birch	$0 \leq M \leq 0.4$	$\varepsilon'(M) = 56.056M^3 - 31.114M^2 + 17.898M + 1.8597$	0.9988
		$\varepsilon''(M) = 0.3628M^2 + 0.0293M - 0.0006$	0.9997
Poplar	$0 \leq M \leq 0.4$	$\varepsilon'(M) = 76.609M^3 - 43.969M^2 + 21.698M + 1.6353$	0.9982
		$\varepsilon''(M) = 0.7202M^2 + 0.0183M - 0.00006$	0.9999
Fir	$0 \leq M \leq 0.4$	$\varepsilon'(M) = 217.61M^3 - 108.03M^2 + 26.77M + 1.5139$	0.9915
		$\varepsilon''(M) = 1.2674M^3 + 0.016M^2 + 0.0284M - 0.0005$	0.9981
Foliage tree	$0 \leq M \leq 0.3$	$\varepsilon'(M) = 137M^3 - 69.688M^2 + 23.385M + 1.4984$	0.9954
	$M > 0.3$	$\varepsilon'(M) = -603.6M^2 + 492.12M - 87.312$	0.9991
	$0 \leq M \leq 0.4$	$\varepsilon''(M) = 1.1541M^3 - 0.5489M^2 + 0.1669M - 0.0004$	0.9937

From: [5].

Table 5.8

Dielectric Properties of Pine Sapwood at 922 MHz

Temperature (°C)	Dielectric Permittivity and Loss Factor	R^2
$20 \leq T \leq 100$	$\varepsilon'(T) = -2 \cdot 10^{-7}T^3 - 0.0015T^2 + 0.2088T + 6.199$	0.9061
	$\varepsilon''(T) = 7 \cdot 10^{-6}T^3 - 0.0016T^2 + 0.0889T + 1.9066$	0.9704

From: [6].

Two thermal parameters of wood—density and thermal conductivity at different temperatures—are given in Table 5.9.

Table 5.9

Density and Thermal Conductivity of Some Wood Samples at Room Temperature

Sample	Temperature (°C)	Density (g/cm^3)	Thermal conductivity (W/m · K)
Birch	20	0.72	0.15
Oak	15	0.825	0.2
Fir	60	0.45	0.11
Cedar	20	0.47	0.095
Maple	30	0.72	0.19
Foliage tree	20	0.6	0.13
Pine ⊥	15	0.545	0.15
Pine //	15	0.545	0.4
Poplar	50	0.58	0.17

From: [7].

Along with wood, diverse fibrous materials such as fabric, leather, paper, and cardboard can be successfully treated by microwaves. Information about the CDP of these materials at ISM frequencies is quite restricted. Experimental results obtained in Khmelnitskiy Technological Institute (Ukraine) for double-component moist fabric at various temperatures are represented in Table 5.10. Table 5.11 contains data about the CDP of four fibrous materials as a function of moisture.

Table 5.10
Dielectric Properties of Polyamide-Cotton Mixture Fabric at 2.45 GHz

W%	$T°C$	15	30	45	60	75	90	105	120
5	ε'	1.2	1.4	2.0	2.4	2.8	3.0	3.5	3.8
	ε''	0.24	0.42	0.7	0.96	1.4	1.8	2.8	3.42
30	ε'	13.8	15.3	16.6	17.4	18.6	20.2	21.4	22.6
	ε''	8.556	6.426	4.98	5.742	4.092	6.06	8.988	13.56

From: [8].

Table 5.11
Dielectric Properties of Some Fibrous Materials at Room Temperature

Sample	W%	Expression	R^2
Paper, 2.45 GHz	$10 \leq W \leq 30$	$\varepsilon'(W) = -0.0003W^3 + 0.0206W^2 - 0.1362W + 3.7$	0.9981
		$\varepsilon''(W) = -3 \cdot 10^{-5}W^3 + 0.0017W^2 + 0.0923W - 0.3$	0.9977
Carton 2.45 GHz	$10 \leq W \leq 30$	$\varepsilon'(W) = 7 \cdot 10^{-5}W^3 - 0.0027W^2 + 0.208W + 1.316$	0.9999
		$\varepsilon''(W) = 0.0622W - 0.192$	0.9887
Leather, 2.45 GHz	$10 \leq W \leq 50$	$\varepsilon'(W) = 3 \cdot 10^{-5}W^3 + 0.0067W^2 - 0.075W + 2.56$	0.9999
		$\varepsilon''(W) = 8 \cdot 10^{-6}W^3 + 0.0038W^2 - 0.0526W + 1.44$	0.9999
Wool, 3 GHz	$4 \leq W \leq 12$	$\varepsilon'(W) = 0.0111W^2 - 0.0401W + 3.692$	0.9981
		$\varepsilon''(W) = 0.0052W^2 - 0.0114W + 0.132$	0.9999

From: [3].

The authors in [9, 10] have developed a mathematical model describing processes of microwave drying of leather. They used the empirical expressions given in Table 5.12 for the modeling variations of the CDP of leather at different levels of moisture.

Finally, in Table 5.13 one can find handbook data on the thermal conductivity of some fibrous materials, taken from [11].

Table 5.12
Dielectric Properties of Leather at 2.45 GHz and
Room Temperature

Expression	Moisture (Dry Basis)	R^2
$\varepsilon'(U) = 12.6919U^2 + 9.0168U + 2.9311$	$0 \le U \le 1.1$	0.9994
$\varepsilon''(U) = 1.511U^2 + 3.8316U + 0.1051$		0.9986

From: [9].

Table 5.13
Thermal Conductivity λ_t (W/m·K) of Some Fibrous Materials

Material	Fiber Diameter (mkm)	$T°C$	λ_{ft} (W/m·K)	Porosity	λ_t
Glass wool	15.5	30	1	0.972	0.044
Asbestos fiber	20	30	1.5	0.8	0.111
Silk	20	30	0.4	0.989	0.037
Flax	18	30	0.5	0.975	0.039
Hemp	20	30	0.5	0.969	0.049
Jute	17	30	0.5	0.887	0.041
Wooden chip	100	30	0.5	0.933	0.051
Felt	37	50	0.3	0.818	0.068
Cotton wool	20	20	0.35	0.947	0.043
Virgin wool	20	20	0.3	0.82	0.052
Kapron	100	20	0.22	0.954	0.033

From: [11].
λ_{ft} = Thermal conductivity of fibers.

References

[1] Torgovnikov, G. I., *Dielectric Properties of Wood and Wood-Based Materials,* Berlin, Germany: Springer-Verlag, 1993.

[2] Tran N., W. K. Tam, and M. Malcmann, "Dielectric Measurements of a Timber Sample under Pressure of Several Bars," *Proceedings of the 4th Congress on Microwave and RF Applications,* Austin, TX, 2004, pp. 416–425.

[3] Metaxas, A. C., and R. J. Meredith, *Industrial Microwave Heating,* London: Peter Peregrinus, 1983.

[4] Olmi, R., M. Bini, and A. Ignesti, et al., "Dielectric Properties of Wood from 2 to 3 GHz," *Journal of Microwave Power and Electromagnetic Energy*, Vol. 35, No. 3, 2000, pp.135–143.

[5] Romanov, A. N., "The Effect of Volume Humidity and the Phase Composition of Water on the Dielectric Properties of Wood at Microwave Frequencies," *Journal of Communication Technology and Electronics*, 2006, Vol. 51, No. 4, pp.435–439.

[6] Tran, N., and M. Malcmann, "The Role of Simulation Software in the Design of Microwave Machine for Heating Large Logs of Timber," *Proceedings of the 9th International Conference on Microwave and High Frequency Heating (AMPERE)*, Loughborough, United Kingdom, 2003, pp. 505–508.

[7] Grigoriev, I. S., and E. Z. Meylihov (eds.), *Physical Constants: Handbook*, Moscow: Energoatomizdat, 1991 (in Russian).

[8] "Analysis of Electrodynamic and Thermal Parameters of Processes of Thermal Treatment of Textile Fabric in Meander-Type Microwave Chamber on Double-Ridged Waveguide," Technical Report No. 112-07/2-92, *Institute of Radio Engineering and Electronics of Russian Academy of Sciences*, Moscow, 1992 (in Russian).

[9] Monzo-Cabrera, J., A. Diaz-Morcillo, and J. M. Catala-Civera, et al., "Effect of Dielectric Properties on Moisture Leveling in Microwave-Assisted Drying of Laminar Materials," *Microwave and Optical Technology Letters*, Vol. 30, No. 3, 2001, pp. 165–168.

[10] Monzo-Cabrera, J., A. Diaz-Morcillo, and J. M. Catala-Civera, et al., "Enthalpy Calculations for the Estimation of Microwave-Assisted Drying Efficiency on Laminar Materials," *Microwave and Optical Technology Letters*, Vol. 31, No. 6, 2001, pp. 470–474.

[11] Dulnev, G. N., and Yu. P. Zarichnyak, *Thermal Conductivity of Mixtures and Composite Materials*, Leningrad: Energia, 1974 (in Russian).

6

Polymers, Resins, and Plastics

According to the classification proposed in [1], resins and plastics are divided into three main groups: (1) no polar high-frequency (HF) dielectrics including polyethylene, polypropylene, polyester and others, (2) weak polar and polar HF and low-frequency (LF) dielectrics including paraformaldehyde, rubbers, polybutadiene, and so forth, and (3) polar LF dielectrics such as polyamide, epoxide, and polyvinylchloride.

Rubber-based resins are multimolecular substances described by the formula $(C_5H_8)m$, where m is the number of molecular chains. Vulcanization (heating of rubber after mixing with sulfur-containing matter) of crude rubber allows improving its heatproof properties. Resins such as polyamide and polyimide have wide practical applications in modern microelectronics and electrical and airspace engineering.

Plastic dielectrics have a weak interaction with electromagnetic (EM) fields because of their nonpolar molecular structure. Plastics are often used as packaging material for foods treated by microwaves. Most plastics have a very weak linear dependence on temperature at RF and microwave frequencies. Table 6.1 lists the CDP values of selected plastic materials at 3 GHz.

Table 6.1
Dielectric Properties of Some Plastic Materials at 3 GHz

Material Labeling	Composition	$T°C$	ε'	ε''
Bakelite BM-120	Phenol-formaldehyde resin 46%, wood flour 46%, additives 8%	25	3.7	0.1628
Formica XX	Phenol-formaldehyde resin 50%, paper 50%, laminate	26	3.57	0.2142
Micarta 254	Cresylic acid formaldehyde 50%, α-cellulose 50%	25	3.43	0.1749
		82	4.02	0.3939
Melmac molding compound 1500	Melamine-formaldehyde resin 42%, wood flour 40%, plasticizer 18%	25	4.2	0.2184
Resimene 803-A	Melamine-formaldehyde resin 60%, cellulose 40%	24	4.53	0.3715

From: [2].

The dielectric properties of various solid and liquid polymers at microwave frequencies are presented in Tables 6.2 to 6.9.

Table 6.2
Dielectric Properties of Some Polyamides at 1 GHz

	$T = 0°C$		$T = 30°C$		$T = 60°C$		$T = 90°C$	
Polymer	ε'	ε''	ε'	ε''	ε'	ε''	ε'	ε''
PA6	3	0.03	3.1	0.062	3.2	0.0832	3.4	0.187
PA66	3	0.0243	3	0.0405	3.1	0.0744	3.2	0.1408
PA610	3	0.03	3	0.039	3	0.075	3.1	0.1085

From: [3].

Table 6.3
Dielectric Properties of Three Polymer Materials at 2.45 GHz

Polymer	Expression	R^2
Resin DER332 [4] $20 \leq T°C \leq 160$	$\varepsilon'(T) = -2 \cdot 10^{-7}T^3 + 0.0001T^2 - 0.001T + 4.4154$	0.9951
	$\varepsilon''(T) = 0.1984exp(0.0125T)$	0.9757
PET [5] $20 \leq T°C \leq 120$	$\varepsilon'(T) = 2.6132exp(0.0034T)$	0.975
	$\varepsilon''(T) = 9 \cdot 10^{-8}T^3 - 0.00001T^2 + 0.0006T + 0.0137$	0.9403
Polyimide [6] $40 \leq T°C \leq 80$	$\varepsilon' = 3.8$	—
	$\varepsilon''(T) = 0.0077ln(T) - 0.0162$	0.9598

Table 6.4

Dielectric Properties of Liquid Monomers in Temperature Range $24 \leq T°C \leq 108$

Monomer	f (MHz)	Dielectric Permittivity and Loss Factor	R^2
ε-Caprolactam	912	$\varepsilon'(T) = -3 \cdot 10^{-7}T^3 - 0.0014T^2 + 0.1757T + 38.814$	0.9483
		$\varepsilon''(T) = -3 \cdot 10^{-7}T^3 + 0.0005T^2 - 0.1809T + 18.146$	0.9952
	2466	$\varepsilon'(T) = -10^{-5}T^3 + 0.0003T^2 + 0.2579T + 21.714$	0.9960
		$\varepsilon''(T) = 10^{-5}T^3 - 0.0033T^2 + 0.1058T + 22.008$	0.9962
ε-Caprolactone	912	$\varepsilon'(T) = -10^{-5}T^3 + 0.0004T^2 + 0.0741T + 40.103$	0.9151
		$\varepsilon''(T) = -6 \cdot 10^{-6}T^3 + 0.0015T^2 - 0.2221T + 17.827$	0.9973
	2466	$\varepsilon'(T) = -8 \cdot 10^{-6}T^3 - 0.0002T^2 + 0.2504T + 23.051$	0.9963
		$\varepsilon''(T) = 6 \cdot 10^{-6}T^3 - 0.0018T^2 + 0.0151T + 22.976$	0.9960

From: [7].

Table 6.5

Dielectric Properties of Four Liquid Polymer
Substances at 2.45 GHz

	DuroLok 270		Stauf Cold Glue 153		ALFO DN 41		NA 62	
$T°C$	ε'	ε''	ε'	ε''	ε'	ε''	ε'	ε''
20	27.1	15.4	23.8	8.8	32.2	13.1	4.19	3.25
30	27.7	16.6	27.7	11.3	31.4	13.1	4.2	3.37
40	27.7	17.1	28.2	12.6	30.8	13.4	4.4	3.41
50	26.7	15.7	28.0	12.7	29.7	13.7	4.7	3.48
60	70.7	15.0	48.0	25	28.4	13.8	4.3	4.11
70	25.7	15.6	23.8	11.4	27.5	15.9	3.8	4.48
80	24.0	16.5	15.1	8.0	26.6	16.5	3.7	3.88

From: [8, 9].

Table 6.6

Dielectric Properties of Wet Movie Film at $T = 25°C$ and 2.97 GHz

W (g/m)	0.1	0.2	0.3	0.4	0.5	0.6	0.7	0.8	0.9	1.0
ε'	14	14.2	21	29	38	47	55	64	72	80
ε''	2.1	4.26	8.4	13.05	15.2	17.39	18.7	20.48	21.6	22.4

From: [10].

Table 6.7
Dielectric Properties of Polymer Resins at 2,450 MHz in
Temperature Range $20 \leq T°C \leq 180$

Polymer Resin	Dielectric Permittivity and Loss Factor	R^2
Sample 2	$\varepsilon'(T) = -0.217 \cdot 10^{-6}T^3 + 0.4525 \cdot 10^{-4}T^2 + 0.0031835T + 2.604$	0.9989
	$\varepsilon''(T) = 0.1594 \cdot 10^{-6}T^3 - 0.7623 \cdot 10^{-4}T^2 + 0.0101337T + 0.22$	0.9693
Sample 3	$\varepsilon'(T) = 0.12816 \cdot 10^{-7}T^3 - 0.6549 \cdot 10^{-4}T^2 + 0.017599T + 2.8094$	0.9865
	$\varepsilon''(T) = 0.6151 \cdot 10^{-6}T^3 - 0.239 \cdot 10^{-3}T^2 + 0.025805T + 0.23564$	0.9781
Sample 4	$\varepsilon'(T) = 0.8 \cdot 10^{-6}T^3 - 0.477 \cdot 10^{-3}T^2 + 0.069575T + 3.69509$	0.9661
	$\varepsilon''(T) = 0.14856 \cdot 10^{-5}T^3 - 0.0005118T^2 + 0.03991T + 1.4116$	0.9972
Sample 5	$\varepsilon'(T) = -0.5978 \cdot 10^{-6}T^3 + 0.10873 \cdot 10^{-3}T^2 + 0.011145T + 3.5315$	0.9831
	$\varepsilon''(T) = 0.2634 \cdot 10^{-6}T^3 - 0.1485 \cdot 10^{-3}T^2 + 0.021547T + 0.4807$	0.9824

From: [11].

Table 6.8
Dielectric Properties of Nylon 6

$T°C$	20	50	75	100	125	150
ε'	3.0	3.1	3.2	3.3	3.7	3.6
ε''	0.02	0.03	0.05	0.07	0.22	0.15

From: [12].

Table 6.9
Dielectric Properties of Crude and Vulcanized Rubber NBR at 2.45 GHz
in Temperature Range $40 \leq T°C \leq 160$

Rubber	Dielectric permittivity and loss factor	R^2
Crude	$\varepsilon'(T) = 2.4667T + 0.0051$	0.9928
	$\varepsilon''(T) = -5 \cdot 10^{-7}T^3 + 7 \cdot 10^{-5}T^2 + 0.0111T - 0.222$	0.9989
Vulcanized	$\varepsilon'(T) = 2 \cdot 10^{-5}T^2 + 0.0016T + 2.5163$	0.9981
	$\varepsilon''(T) = -6 \cdot 10^{-5}T^2 + 0.0219T - 0.6298$	0.9878

From: [13].

Thermal properties of polymers have been also reported in many publications. For example, temperature dependencies of thermal conductivity (λ_t,W/m·K), density (ρ_t, kg/m^3) and heat capacity (C_t, kJ/kg·K) of high-density polyethylene (HDPE), low-density polyethylene (LDPE), and polyvinylchloride (PVC) [14]:

$$\lambda_t(T) = a_1 \cdot 10^{-1} + a_2 \cdot 10^{-4}(T - 273.2) + a_3 \cdot 10^{-6}(T - 273.2)^2$$
$$+ a_4 \cdot 10^{-8}(T - 273.2)^3 + a_5 \cdot 10^{-10}(T - 273.2)^4 \qquad (6.1)$$

$$\left(\rho_t(T)\right)^{-1} = b_1 + b_2 \cdot 10^4(T - 273.2) + b_3 \cdot 10^{-5}(T - 273.2)^2$$
$$+ b_4 10^{-7}(T - 273.2)^3 + b_5 \cdot 10^{-4}(T - 273.2)^4 + b_6 \cdot 10^{-11}(T - 273.2)^5 \qquad (6.2)$$

$$C_t(T) = d_1 + d_2(T - 273.2) + d_3(T - 273.2)^2$$
$$+ d_4(T - 273.2)^3 + d_5(T - 273.2)^4 \qquad (6.3)$$

where a_i, b_i, d_i are the coefficients given in Tables 6.10 to 6.12.

Table 6.10
Temperature Ranges and Coefficients of (6.1)

Polymer	Temperature (°K)	a_1	a_2	a_3	a_4	a_5
HDPE	$283 \leq T < 416$	4.53	−8.59	−5.29	4.12	−1.98
	$416 \leq T \leq 473$	2.6	0			
LDPE	$283 \leq T < 399$	3.65	−4.07	−7.34	8.28	−5.53
	$399 \leq T \leq 473$	2.23	0			
PVC	$273 \leq T \leq 473$	1.68	0			

Table 6.11

Temperature Ranges and Coefficients of (6.2)

Polymer	Temperature (°K)	b_1	b_2	b_3	b_4	b_5	b_6
HDPE	$283 \leq T < 406$	1.033	17.87	−7.19	16.11	−15.45	5.58
	$406 \leq T \leq 473$	1.158	8.09	0			
LDPE	$283 \leq T < 406.5$	1.078	1.24	2.68	−3.95	2.35	0
	$406.5 \leq T \leq 473$	1.158	8.09	0			
PVC	$283 \leq T < 383$	0.7154	1.02	0.0781	−0.0167	0.0524	0
	$383 \leq T \leq 473$	0.6791	5.67	0			

Table 6.12

Temperature Ranges and Coefficients of (6.3)

Polymer	Temperature (°K)	d_1	d_2	d_3	d_4	d_5
HDPE	$283 \leq T < 361$	1.597	$3.61 \cdot 10^{-3}$	$5.96 \cdot 10^{-5}$	$-3.44 \cdot 10^{-8}$	$9.77 \cdot 10^{-9}$
	$361 \leq T \leq 394$	−198.3	6.17	−0.0634	$2.19 \cdot 10^{-4}$	0
	$394 \leq T \leq 403$	−283.7	2.41	0		
	$403 \leq T \leq 406$	1,208	−9.07	0		
	$406 \leq T \leq 473$	1.984	$3.88 \cdot 10^{-3}$	0		
LDPE	$283 \leq T < 363$	1.943	$5.39 \cdot 10^{-3}$	$2.56 \cdot 10^{-2}$	$-3.23 \cdot 10^{-6}$	$3.53 \cdot 10^{-8}$
	$363 \leq T \leq 378$	84.97	−1.84	0.0104	0	
	$378 \leq T \leq 383$	−129	1.3	0		
	$383 \leq T \leq 386$	378.6	−3.31	0		
	$386 \leq T \leq 473$	1.98	$3.7 \cdot 10^{-3}$	0		
PVC	$283 \leq T < 340$	0.75	$4.66 \cdot 10^{-3}$	0		
	$340 \leq T \leq 369$	136.1	−6.64	0.121	$-9.71 \cdot 10^{-4}$	$2.9 \cdot 10^{-6}$
	$369 \leq T \leq 473$	1.208	$2.96 \cdot 10^{-3}$	0		

In many practical cases, information about fixed values of thermal parameters of polymers at room temperature can be also useful (see Table 6.13).

Table 6.13
Thermal Properties of Some Polymers at Room Temperature

Polymer	Thermal Conductivity (W/(m·K))	Heat Capacity (J/(g·K))	Density (kg/m³)
Polypropylene	0.172	2.14	0.85
Polyisobutyene	0.13	1.97	0.86
Polystyrene	0.142	1.21	1.05
Poly(vinyl chloride)	0.168	0.96	1.39
Poly(vinyl acetate)	0.159	1.47	1.19
Poly(vinyl carbazole)	0.155	1.26	1.19
Poly(methyl methacrylate)	0.193	1.38	1.17
Polyisoprene	0.131	1.89	0.91
Polychloroprene	0.193	1.59	1.24
Poly(ethylene oxide)	0.205	2.01	1.13
Polyurethane	0.147	1.7	1.05
Poly(dimethyl siloxane)	0.163	1.59	0.98
Phenolic resin	0.176	1.05	1.22
Epoxide resin	0.18	1.25	1.19
Polyester resin	0.176	1.25	1.23

From: [15].

Finally, in Table 6.14 one can find approximate expressions for two parameters as a function of temperature for two different polymers.

Table 6.14
Heat Capacity (kJ/kg·K) and Density (kg/m3) of Two Polymers

Polypropylene [16]	$C_t(T) = 1.3929\exp(0.0052T)$, $0 \leq T°C \leq 125$	$R^2 = 0.9964$
Poly(vinyl chloride) [3]	$\rho_t(T) = -0.0037T^2 - 0.0104T + 1393.6$, $20 \leq T°C \leq 97$	$R^2 = 0.9995$

References

[1] Kazarnovskiy, D. M., and S. A. Yamanov, *Radio Engineering Materials*, Moscow: Vishaya Shkola Issue, 1972 (in Russian).

[2] Von Hippel, A. R., *Dielectric Materials and Applications*, New York: MIT Press, 1954.

[3]   Brandrup, J., E. H. Immergut, and E. A. Grulke (eds.)., *Polymer Handbook*, Hoboken, NJ: John Wiley and Sons, Inc., 1999.

[4]   Jow J., M. C. Hawley, and M. Finzel, et al., "Microwave Processing and Diagnostics of Chemically Reacting Materials in a Single Mode Cavity Applicator," *IEEE Transactions on Microwave Theory and Techniques*, Vol. 35, No. 12, 1987, pp. 1435–1443.

[5]   Estel L., A. Ledoux, and C. Bonnet, "Microwave Assisted Blow Molding of Polyethyleneterephthalate (PET) Bottles," *Proceedings of the 4th World Congress on Microwave and Radio Frequency Applications*, Austin, TX, 2004, pp. 220–231.

[6]   Nikawa, Y. and Y. Kanai, "Measurement of Complex Permittivity and Temperature Dependence for Thin Material within Cavity Resonator," *Proceedings of the 40th International Microwave Power Symposium*, Boston, 2006, pp. 82–84.

[7]   Hutcheon, R., J. Mouris, and X. Fang, et al., "Measurements of the High-Temperature Microwave (400-3000 MHz) Complex Dielectric Constants of Monomers ε-Caprolactam and ε-Caprolactone," *Microwaves: Theory and Application in Materials Processing V*, Proceedings of the 2nd World Congress on Microwave and RF Processing, 2000, pp.109–118.

[8]   Abbas, M., P. A. Bernard, and C. Marzat, et al., "Complex Permittivity Measurements and Electromagnetic Field Modeling: Application to Element Assembly by Microwave Assisted Gluing," *Proceedings of the 5th International Conference on Microwave and High Frequency Heating*, Cambridge, United Kingdom, 1995, pp. 23.1–24.4.

[9]   Kayser, T., M. Pauli, and W. Sorgel, et al., "Investigations and Case Studies of Microwave Heating in Parquet Industry," *Proceedings of the 4th Congress on Microwave and RF Applications*, Austin, TX, 2004, pp. 67–74.

[10]   Komarov, V. V., and V. V. Yakovlev, "Modeling over Determination of Dielectric Properties by the Perturbation Technique," *Microwave and Optical Technology Letters*, Vol. 39, No. 6, 2003, pp. 443–446.

[11]   Akhtar, M. J., L. Feher, and M. Thumm, "A Novel Approach for Measurement of Temperature Dependent Dielectric Properties of Polymer Resins at 2.45 GHz," *Proceedings of the First Global Congress on Microwave Energy Applications*, Otsu, Japan, August 2008, pp. 529–532.

[12]   Liu, F., I. Turner, and E. Siores, et al., "A Numerical and Experimental Investigation of the Microwave Heating of Polymer Materials Inside a Ridge Waveguide," *Journal of Microwave Power and Electromagnetic Energy*, Vol. 31, No. 2, 1996, pp. 71–82.

[13]   Catala-Civer, J. M., S. Giner-Maravilla, and D. Sanchez-Hernandez, et al., "Pressure-Aided Microwave Rubber Vulcanization in a Ridged Three-Zone Cylindrical Cavity," *Journal of Microwave Power and Electromagnetic Energy*, Vol. 35, No. 2, 2000, pp. 92–104.

[14]   Bodling, B. A. and M. Prescott (eds.), *Heat Exchanger Design Handbook*, Vol. 5, New York: Hemisphere Publishing Corp., 1983.

[15]   Van Krevelin, D.W. (ed.), *Properties of Polymers*, New York: Elsevier Inc. 1990.

[16]   Maier, C., and T. Calafat, *Polypropylene: The Definitive User's Guide and Databook*, Norwich, NY: Plastic Design Library, Inc., 1998.

7

Ceramics

Ceramics are widely applied in many areas of science and engineering because of unique features such as high-temperature stability of physical properties, low wearability, thermal conductivity, and weight. Dielectric properties of selected ceramics as a function of temperature and frequency have been studied in several publications [1–5]. The dielectric constant of ceramics may be sufficiently low ($\varepsilon' < 10$); for example, for Mullite, average ($30 \leq \varepsilon' \leq 150$) for titanic ceramics N1400-110, N750T in frequency range 10 Hz $\leq f \leq 10^4$ MHz or high ($\varepsilon' > 200$) for TiO_2, $CaTiO_3$, and $LaAlO_3$.

The complex dielectric permittivity of the ceramic Hilox-882, which consists of Al_2O_3, SiO_2, MgO, CrO_2, and CaO at two microwave frequencies as a function of temperature is shown in Table 7.1 [1].

Table 7.1
Dielectric Properties of Hilox-882 in Temperature Range $20 \leq T°C \leq 1{,}200$

Frequency	Dielectric Permittivity and Loss Factor	R^2
915 MHz	$\varepsilon'(T) = -6 \cdot 10^{-11}T^3 + 3 \cdot 10^{-6}T^2 - 0.0005T + 8.559$	0.9947
	$\varepsilon''(T) = 0.0951exp(0.0032T)$	0.9767
2,450 MHz	$\varepsilon'(T) = 7 \cdot 10^{-10}T^3 + 2 \cdot 10^{-6}T^2 - 0.0004T + 8.849$	0.9966
	$\varepsilon''(T) = 3 \cdot 10^{-9}T^3 - 3 \cdot 10^{-6}T^2 + 0.0012T - 0.0605$	0.9893

From: [1].

The analytical model that describes the behavior of ε' and ε'' of composite ceramics in temperature range $0 < T°C < 2{,}500$ at 2,450 MHz has been derived in [2]:

$$\varepsilon = \varepsilon_{in} + 10^{-4}\left(T - 25\right) - j10^{-3}\left(T - 25\right) \tag{7.1}$$

where $\varepsilon_{in} = 3.9 - j0.46$ is the initial value at $T = 25°C$.

Several types of ceramics—aluminosilicate and zirconia fiber board, alumina, Ni-Zn-ferrite, and PZT ceramics—have been analyzed at 2,422 MHz in [3], where these materials are treated by microwave in a cylindrical applicator with operating mode TM_{010}.

Measured values of $\varepsilon'(T) - j\varepsilon''(T)$ at 3,000 MHz for soda-lime glass (Corning 0080), borosilicate glass, Mullite MV20, cupric oxide, aluminum nitride, zeolite, alumina cement AC56, zirconia cement, and felt can be found in [4] (see Table 7.3).

Some of these dependencies (zeolite) may be approximated by expressions:

$$\varepsilon' = 0.250933T^{0.499713}, \quad \varepsilon'' = 0.383887\exp\left(0.0047014T\right) \quad R^2 = 0.962 \tag{7.2}$$

Both equations are valid when $300 < T°C < 500$.

Table 7.2
Dielectric Properties of Some Ceramic Materials at 2.422 GHz

Material	Temperature	Expression	R^2
Aluminosilicate fiberboard	$20 \leq T°C \leq 1300$	$\varepsilon'(T) = 2\cdot 10^{-10}T^3 - 6\cdot 10^{-8}T^2 - 0.0003T + 1.2671$	0.8933
		$\varepsilon''(T) = 2\cdot 10^{-11}T^3 + 8\cdot 10^{-9}T^2 - 9\cdot 10^{-6}T + 0.0027$	0.9940
Zirconia fiber board	$20 \leq T°C \leq 1400$	$\varepsilon'(T) = 6\cdot 10^{-10}T^3 + 9\cdot 10^{-7}T^2 - 0.0003T + 2.5433$	0.9931
		$\varepsilon''(T) = 3\cdot 10^{-9}T^3 - 5\cdot 10^{-7}T^2 - 0.0003T + 0.0438$	0.9995
Aluminosilicate saggar	$20 \leq T°C \leq 1350$	$\varepsilon'(T) = 7\cdot 10^{-10}T^3 - 9\cdot 10^{-7}T^2 + 0.0008T + 4.0371$	0.9962
		$\varepsilon''(T) = 3\cdot 10^{-10}T^3 - 3\cdot 10^{-7}T^2 + 0.0001T - 0.0028$	0.9916
PZT	$20 \leq T°C \leq 320$	$\varepsilon'(T) = 0.0005T^2 + 0.0037T + 34.062$	0.9945
	$400 \leq T°C \leq 999$	$\varepsilon'(T) = 10^6 T^{-1.6093}$	0.9993

From: [3].

Table 7.3
Dielectric Properties of Some Ceramic Materials at 3 GHz

Material	Temperature	Expression	R^2
Corning 0080	$300 \leq T°C \leq 750$	$\varepsilon'(T) = 4 \cdot 10^{-7}T^3 - 0.0005T^2 + 0.2162T - 22.495$	0.9977
		$\varepsilon''(T) = 0.0122exp(0.0089T)$	0.9246
Mullite	$500 \leq T°C \leq 1300$	$\varepsilon'(T) = 5.3076exp(0.0003T)$	0.9757
		$\varepsilon''(T) = 3 \cdot 10^{-9}T^3 - 7 \cdot 10^{-6}T^2 + 0.0059T - 1.5404$	0.9978
Corning 7740	$300 \leq T°C \leq 900$	$\varepsilon'(T) = 3.5197exp(0.0007T)$	0.8372
		$\varepsilon''(T) = 5 \cdot 10^{-8}T^3 - 7 \cdot 10^{-5}T^2 + 0.0319T - 4.7674$	0.9920
Corning 3320	$300 \leq T°C \leq 950$	$\varepsilon'(T) = 4 \cdot 10^{-8}T^3 - 6 \cdot 10^{-5}T^2 + 0.0295T - 0.0432$	0.9968
		$\varepsilon''(T) = 4 \cdot 10^{-8}T^3 - 6 \cdot 10^{-5}T^2 + 0.0302T - 4.784$	0.9925
Al-N powder	$350 \leq T°C \leq 1150$	$\varepsilon'(T) = 3 \cdot 10^{-6}T^2 - 0.0012T + 3.4395$	0.9799
		$\varepsilon''(T) = 2 \cdot 10^{-6}T^2 - 0.0008T - 0.0132$	0.9807
Cupric oxide	$300 \leq T°C \leq 900$	$\varepsilon'(T) = 3.1059exp(0.0013T)$	0.9805
		$\varepsilon''(T) = 24.942ln(T) - 135.86$	0.9016
Linde 13X	$300 \leq T°C \leq 500$	$\varepsilon'(T) = 2.9677exp(0.0013T)$	0.9644
		$\varepsilon''(T) = 0.3865exp(0.0047T)$	0.9298
Zircar ZC89	$300 \leq T°C \leq 950$	$\varepsilon'(T) = 8 \cdot 10^{-9}T^3 - 10^{-5}T^2 + 0.0036T + 2.7885$	0.9957
		$\varepsilon''(T) = 4 \cdot 10^{-9}T^3 - 3 \cdot 10^{-6}T^2 + 0.0004T + 0.1906$	0.9915
Zircar AC56	$300 \leq T°C \leq 1100$	$\varepsilon'(T) = 4 \cdot 10^{-9}T^3 - 10^{-5}T^2 + 0.0095T + 2.2372$	0.9869
		$\varepsilon''(T) = -9 \cdot 10^{-11}T^3 + 4 \cdot 10^{-7}T^2 - 0.0002T + 0.0206$	0.9895
Zircar ZYF100	$300 \leq T°C \leq 1000$	$\varepsilon'(T) = -3 \cdot 10^{-10}T^3 - 8 \cdot 10^{-8}T^2 + 0.0014T + 1.3788$	0.9881
		$\varepsilon''(T) = -3 \cdot 10^{-9}T^3 + 7 \cdot 10^{-6}T^2 - 0.003T + 0.3662$	0.9950

From: [4].

The ISM frequency 915 MHz is often utilized for ceramic processing. Information about $\varepsilon'(T) - j\varepsilon''(T)$ of two ceramic materials at this frequency and high temperatures is given in Table 7.4.

Table 7.4
Dielectric Properties of Ceramics at 915 MHz

Ceramics	Temperature Range	Expression	R^2
Mullite	$400 \leq T°C \leq 1300$	$\varepsilon'(T) = 3 \cdot 10^{-9}T^3 - 4 \cdot 10^{-6}T^2 + 0.0032T + 4.4869$	0.9935
		$\varepsilon''(T) = 9 \cdot 10^{-7}T^2 - 0.0002T - 0.0268$	0.9997
Silicon carbide powder	$500 \leq T°C \leq 950$	$\varepsilon'(T) = 21.051exp(0.0004T)$	0.9274
		$\varepsilon''(T) = 1.913exp(0.0015T)$	0.9566

From: [5].

Sometimes the physical properties of ceramic materials strongly depend on the surrounding atmosphere, as shown in Table 7.5 for the ZnO sample with $\varepsilon'(T) \approx 4$ and $\varepsilon''(T) = var$.

Table 7.5
Loss Factor of Ceramics ZnO in at 2.45 GHz

$T°C$	50	100	150	200	250	300	350	400	450	500	550	600
Air Atmosphere												
ε''	0.13	0.145	0.16	0.26	0.21	0.13	0.1	0.099	0.1	0.105	0.116	0.14
Nitrogen Atmosphere												
ε''	0.7	1.3	2.0	3.9	8.1	5.8	5.3	14.6	23.2	7.7	0.9	0.2

From: [6].

The CDP of the ceramic ZnO has been also studied in [7] as a function of temperature for two frequencies (see Table 7.6).

Table 7.6
Dielectric Properties of ZnO

	615 MHz		**2,214 MHz**	
$T°C$	ε'	ε''	ε'	ε''
20	1.9	0.005	1.93	0.004
100	1.86	0.02	1.83	0.007
200	1.9	0.03	1.86	0.009
300	1.93	0.04	1.9	0.011
400	1.965	0.05	1.93	0.013
500	2	0.06	1.965	0.015
600	2.1	0.07	1.98	0.017
700	2.207	0.08	2	0.023
800	2.31	0.09	2.07	0.035
900	2.724	0.04	2.276	0.168
1,000	3.793	1.356	3.034	0.62
1,500	3.586	1.834	2.58	0.05

From: [7].

Two ceramic materials have been studied in [8] and proposed for application in microwave tubes. The CDP values of these materials are given in Table 7.7.

Table 7.7
Dielectric Properties of Ceramic Materials at 3,340 MHz and $0 \leq T°C \leq 500$

Ceramic	Complex Dielectric Permittivity	R^2
22XC	$\varepsilon'(T) = 0.0011T + 9.3095$	0.9985
	$\varepsilon''(T) = 1.0556 \cdot 10^{-10}T^3 + 1.119 \cdot 10^{-8}T^2 + 1.31 \cdot 10^{-5}T + 0.0075047$	0.9991
A995	$\varepsilon'(T) = 0.0009T + 10.092$	0.9951
	$\varepsilon''(T) = 3 \cdot 10^{-11}T^3 - 10^{-8}T^2 + 5 \cdot 10^{-6}T + 0.0001$	0.9945

From: [8].

The dielectric properties of alkali metal (K_2CO_3, Na_2CO_3, and Li_2CO_3) and alkali earth carbonates ($BaCO_3$, $CaCO_3$, $SrCO_3$, and $MgCO_3$) at 2.45 GHz in temperature interval $20 \leq T°C \leq 1,100$ can be found in [9]. The dielectric properties of barium carbonate at 2.45 GHz in temperature range $50 \leq T°C \leq 800$ [10]:

$$\varepsilon'(T) = 10^{-8}T^3 - 10^{-5}T^2 + 0.0034T + 2.863, \ R^2 = 0.9898 \qquad (7.3)$$

$$\varepsilon''(T) = 7 \cdot 10^{-11}T^3 - 4 \cdot 10^{-8}T^2 + 3 \cdot 10^{-6}T + 0.0017, \ R^2 = 0.9811 \qquad (7.4)$$

Usually scientific publications only contain data about the CDP or some thermal parameters of ceramics, and information about both dielectric and thermal parameters of one material is quite rare in the literature. Table 7.8 lists such a case when the four main parameters of the ceramic mullite with density $\rho_t = 2500 \ \text{kg/m}^3$ in the temperature range $27 \leq T°C \leq 1077$ are represented.

Table 7.8
Dielectric and Thermal Properties of Mullite at 2.45 GHz

Parameter	R^2
$\varepsilon'(T) = 2.119 \cdot 10^{-6}T^2 - 0.000337T + 6.1438$	0.9994
$\varepsilon''(T) = 1.7052 \cdot 10^{-9}T^3 - 1.4616 \cdot 10^{-6}T^2 + 0.000559T + 0.02279$	0.9986
$\lambda_t(T) = 2.643 \cdot 10^{-6}T^2 - 0.005067T + 5.83152$	0.9977
$C_t(T) = 152.22\ln(T) + 199.5$	0.9989

From: [11].

Thermal parameters of ceramics and glasses are usually measured at high temperatures because these materials are obtained by sintering of inorganic substances, including minerals and oxides. Information about density $\rho_t(T)$, heat capacity $C_t(T)$, and thermal conductivity $\lambda_t(T)$ of different ceramics and glasses is available in [11–21]. Some of this data is represented in Tables 7.9–7.20.

Table 7.9
Thermal Conductivity λ_t (W/m·K) of Porous and Solid Ceramics

p_r,%	$T°C$	100	200	400	600	800	1000	1200	1400	1600
4.5 ÷ 7.3	Al_2O_3	28.84	21.28	12.56	8.72	6.86	5.85	5.28	5.23	5.78
0		30.24	22.44	13.14	9.13	7.21	6.14	5.52	5.50	6.07
4.7 ÷ 9.9	BeO	209.3	166.3	88.39	44.89	25.76	19.31	16.40	15,58	14.54
0		219.8	174.5	93.04	46.98	26.98	20.24	17.21	16.39	15.23
3 ÷ 8	MgO	34.42	26.98	15.82	11.03	8.141	6.687	5.873	5.78	6.570
0		36.05	28.26	16.51	11.51	8.489	6.989	6.106	6.048	6.86
36	Mg_2SiO_4	36.86	30.94	2.465	2.058	1.873	1.674	1.628	1.593	—
0		5.373	4.512	3.558	2.977	2.675	2.43	2.372	2.30	—
13	ZrO_2	1.674	1.721	1.756	1.802	1.88	1.965	2.047	2.09	—
0		1.954	1.965	2.046	2.093	2.198	2.279	2.396	2.44	—
3.5	TiO_2	6.28	4.826	3.744	3.465	3.270	3.186	3.186	—	—
0		6.53	5.01	3.90	3.628	3.396	3.303	3.30	—	—
30	Graphite	124.78	101.8	78.74	64.54	53.49	43.96	38.49	—	—
0		177.9	144.8	112.22	92.22	76.18	62.33	54.66	—	—
8.75	CaO	13.96	10.12	8.37	7.582	7.28	7.117	—	—	—
0		15.235	11.107	9.188	8.292	7.978	7.792	—	—	—
18.6	$ZrSiO_4$	—	4.628	4.187	3.768	3.490	3.303	3.187	3.094	—
0		—	5.699	5.187	4.652	4.303	4.093	3.93	3.84	—

From: [12].
p_r = porosity.

Table 7.10
Thermal Conductivity λ_t (W/m·K) and Heat Capacity C_t(J/kg·K) of Glasses

Glass Type	Temperature Range	Expression	R^2
Organic	$7 \leq T°C \leq 97$	$\lambda_t(T) = 0.0031ln(T) + 0.1863$	0,9762
		$C_t(T) = 4.2455T + 1274.5$	0,9993
Quartz	$27 \leq T°C \leq 387$	$\lambda_t(T) = -2 \cdot 10^{-9}T^3 + 6 \cdot 10^{-7}T^2 + 0.0014T + 1.3031$	0,9968
		$C_t(T) = 7 \cdot 10^{-7}T^3 - 0.0019T^2 + 1.614T + 696.28$	0,9999
Optical	$27 \leq T°C \leq 227$	$\lambda_t(T) = 3 \cdot 10^{-6}T^2 + 0.0003T + 0.6921$	0,9977
		$C_t(T) = -0.0018T^2 + 1.0271T + 441.01$	0,9998

From: [13].

Table 7.11
Thermal Properties of Various Type Bricks

Brick Type	$T°C$	ρ_t (kg/m³)	λ_t (W/m·K)	C_t (J/kg·K)
Fire	393	1,890	1.093	1,004.8
	598	1,890	1.1514	1,067.6
	791	1,890	1.198	1,084
Magnesium	100	2,350	5.815	1,047
	1000	2,350	3.49	1,423
Carborundum	100	2,300	9.3	837
	1,000	2,300	11.05	1,256

From: [13].

Heat capacity (C_t, J/kg·K) of SiC fiber-reinforced SiC-matrix composite in temperature range $25 \leq T°C \leq 700$ [14]:

$$C_t = 2 \cdot 10^{-6} T^3 - 0.0032 T^2 + 2.1114 T + 613.71, \; R^2 = 0.9988 \tag{7.5}$$

Table 7.12
Thermal Conductivity (λ_t, W/m·K) and Diffusitivity
(a_t 10^{-6}, m²/s) of CVI Ceramics

$T°C$	100	250	500	750	1,000	1,150	1,300
λ_t, W/(m·K)	1.3	1.27	1.24	1.7	2	2.5	3.3
$a_t \cdot 10^{-6}$, m²/s	0.8	0.7	0.6	0.75	0.9	0.95	1.0

From: [15].

The thermal conductivity (λ_t, W/m·K) of boron nitride ($O_2 = 1\%$, $\rho_t = 0.45$ g/cm³) in temperature range $20 \leq T°C \leq 280$ [16]:

$$\lambda_t(T) = 2 \cdot 10^{-7} T^3 - 5 \cdot 10^{-5} T^2 - 0.0131 T + 8.6579, \; R^2 = 0.9944 \tag{7.6}$$

Table 7.13

Thermal Properties of Ceramics in Temperature Range $300 \leq T°C \leq 1{,}000$

Ceramic	Parameter	Expression	R^2
Si_3N_4 + AlN + Si-Al-O-N	C_t, J/kg·K	$C_t(T) = -0.0012T^2 + 2.1487T + 228.34$	0,9955
	λ_t, W/m·K	$\lambda_t(T) = -5.3585 \cdot 10^{-5}T + 5.668$	0,9999
	a_t, cm^2/s	$a_t(T) = 0.0318exp(-0.0007T)$	0.9582
Si_3N_4 + MgO	C_t, J/kg·K	$C_t(T) = -0.0013T^2 + 2.2514T + 108.75$	0,9703
	λ_t, W/m·K	$\lambda_t(T) = 76.962exp(-0.0011T)$	0,9853
	a_t, cm^2/s	$a_t(T) = -3.35714 \cdot 10^{-4}T + 0.4007$	0.9999

From: [17].

Table 7.14

Density $(\rho_t, g/cm^3)$ of Infrasil-Silica and Amersil-Silica Glass

Glass	Infrasil Silica				Amersil Silica				
$T°C$	1,000	1,100	1,200	1,400	1,935	2,048	2,114	2,165	2,322
ρ_t, g/cm^3	2.201	2.198	2.206	2.213	2.094	2.072	2.057	2.045	1.929

From: [18].

Table 7.15

Thermal Properties of Glasses

Glass	Temperature	Expression	Units	R^2
Li_2O + SiO_2	$20 \leq T°C \leq 1450$	$\rho_t(T) = 2.48 - 0.0004T$	g/cm^3	0.9998
Na_2O + SiO_2		$\rho_t(T) = -0.0012T^2 + 2.1487T + 228.34$		0.9957
KV silica	$80 \leq T°C \leq 700$	$\lambda_t(T) = 0.5712ln(T) - 1.9281$	W/m·K	0.9984
	$97 \leq T°C \leq 470$	$C_t(T) = 172.59ln(T) + 40.293$	J/kg·K	0.9980

From: [18].

Table 7.16
Density (g/cm^3) of Ceramics at Room Temperature

CrB$_2$	HfB$_2$	TaB$_2$	TiB$_2$	ZrB$_2$	B$_4$C	HfC	SiC	TaC	TiC
5.6	11.2	12.6	4.55	12.6	2.51	12.6	3.22	14.57	4.93
Cr$_3$C$_2$	ZrC	AlN	BN	TiN	Si$_3$N$_4$	Al$_2$O$_3$	BeO	CaO	Cr$_2$O$_3$
6.7	6.55	3.28	3.49	5.43	3.18	3.98	3.02	3.32	5.21
MgO	HfO$_2$	NiO	TiO$_2$	UO$_2$	ZrO$_2$	2MgO2Al$_2$O$_3$5SiO$_2$			
3.58	9.68	7.13	4.25	10.96	5.56	1.61 ÷ 2.5			
3Al$_2$O$_3$2SiO$_2$ (mullite)			Al$_2$O$_3$MgO (spinel)			SiO$_2$ZrO$_2$ (zircon)			
2.6 ÷ 3.26			3.58			4.6			

From: [19].

Table 7.17
Thermodynamic Density of Glasses and Ceramics

Ceramic	Density (g/cm^3)	Temperature
B$_2$O$_3$ glass	$\rho_t(T) = 10^{-11}T^3 + 2\cdot10^{-7}T^2 - 0.0005T + 1.8658$, $R^2 = 0.9967$	$25 \leq T°C \leq 1{,}400$
SiO$_2$ glass	$1.93 \leq \rho_t(T) \leq 2.057$	$2{,}114 \leq T°C \leq 2{,}322$
B$_2$O$_3$-CaO, 37.1% CaO	$2.26 \leq \rho_t(T) \leq 2.82$	$25 \leq T°C \leq 1{,}200$
SiO$_2$-Na$_2$O, 20% NaO	$2.22 \leq \rho_t(T) \leq 2.27$	$987 \leq T°C \leq 1{,}388$
SiO$_2$-Na$_2$O, 60% NaO	$2.145 \leq \rho_t(T) \leq 2.25$	$1{,}052 \leq T°C \leq 1{,}413$
SiO$_2$-Al$_2$O$_3$, 14.8% Al$_2$O$_3$	$2.302 \leq \rho_t(T) \leq 2.32$	$1{,}707 \leq T°C \leq 2{,}008$
SiO$_2$-B$_2$O$_3$, 53.1% B$_2$O$_3$	$\rho_t(T) = 1.892 - 0.0634\cdot10^{-3}T$	$1{,}653 \leq T°K \leq 1{,}803$

From: [19].

The heat capacity (J/g·K) of α-alumina ceramic [20]:

$$C_t(T) = \sum_{n=0}^{7} q_n T^n \qquad (7.7)$$

Table 7.18
Coefficients of (7.7)

q_0	q_1	q_2	q_3
−0.581126	$8.25981 \cdot 10^{-3}$	$-1.76767 \cdot 10^{-5}$	$2.17663 \cdot 10^{-8}$
q_4	q_5	q_6	q_7
$-1.60541 \cdot 10^{-11}$	$7.01732 \cdot 10^{-15}$	$-1.67621 \cdot 10^{-18}$	$1.68486 \cdot 10^{-22}$

Table 7.19
Density and Heat Capacity of Ceramics

Ceramic Composition %	Density (g/cm³)	Heat capacity (kJ/kg·K)	
		$T = 300°C$	$T = 1200 °C$
93...96 SiO_2	$1.7 \leq \rho_t(T) \leq 1.9$	0.92	1.13
15...25 Al_2O_3, 75...85 SiO_2	$1.7 \leq \rho_t(T) \leq 2.1$	0.88	1.05
60...72 Al_2O_3, 28...40 SiO_2	$2.2 \leq \rho_t(T) \leq 2.4$	0.92	1
73...75 Al_2O_3, 18...20 SiO_2	$2.9 \leq \rho_t(T) \leq 3$	0.24	0.26
80...99 Al_2O_3	$2.5 \leq \rho_t(T) \leq 3.2$	0.92	1.13
80...95 MgO, Rest: Fe_2O_3, Al_2O_3	$2.6 \leq \rho_t(T) \leq 3.1$	0.24	0.29
47...55 MgO, 33...39 SiO_2 0...11 Fe_2O_3	$2.4 \leq \rho_t(T) \leq 2.6$	0.88	1
66 ZrO_2, 33 SiO_2	$3.2 \leq \rho_t(T) \leq 3.5$	0.63	0.75
50...95 SiC, Rest: Al_2O_3, SiO_2	$2.2 \leq \rho_t(T) \leq 2.7$	8.4	1.13
90...98 C	$1.3 \leq \rho_t(T) \leq 1.8$	1.25	1.59

From: [21].

Table 7.20
Heat Capacity of Some Ceramics and Glasses

Material	Temperature	C_t [kJ/(kg·K)]
High-frequency china	300°K	0.75
Low-frequency china	300°K	0.85
Zirkon	273°K	0.55
	1273°K	0.68
Quartz glass	293°K	0.89
	873°K	1
	1473°K	1.14

From: [22].

Table 7.21
Thermal Conductivity (W/m · K) of Ceramics

Ceramics	Temperature	Expression	R^2
Corundum	$200 \leq T°C \leq 1{,}400$	$\lambda_t(T) = 10^{-6}T^2 - 0.0049T + 8.38$	0.9999
Schamotter, $\rho_t = 1750$ kg/m^3	$200 \leq T°C \leq 1{,}200$	$\lambda_t(T) = 0.0004T + 0.514$	0.9997
C90%, $\rho_t = 1{,}190$ kg/m^3	$200 \leq T°C \leq 1{,}400$	$\lambda_t(T) = 4 \cdot 10^{-7}T^2 + 0.0006T + 0.78$	0.9999
Graphite	$200 \leq T°C < 800$	$\lambda_t(T) = -0.005T + 128$	0.9999
	$800 < T°C \leq 1{,}400$	$\lambda_t(T) = -0.01T + 132$	0.9999
SiC, $\rho_t = 3{,}217$ kg/m^3	$0 \leq T°C \leq 1{,}200$	$\lambda_t(T) = 0.7773 exp(0.0009T)$	0.9997
Magnesite 95%	$0 \leq T°C \leq 1{,}400$	$\lambda_t(T) = -6 \cdot 10^{-9}T^3 + 2 \cdot 10^{-5}T^2 - 0.0251T + 15.801$	0.9999
CrMgO	$0 \leq T°C \leq 1{,}400$	$\lambda_t(T) = 3 \cdot 10^{-6}T^2 - 0.0092T + 9.6002$	0.9999
Sillimanite, $\rho_t = 2450$ kg/m^3	$0 \leq T°C \leq 1{,}400$	$\lambda_t(T) = 8 \cdot 10^{-8}T^2 - 0.0004T + 1.7098$	0.9995
Sillimanite, $\rho_t = 1100$ kg/m^3	$0 \leq T°C \leq 1{,}400$	$\lambda_t(T) = 10^{-7}T^2 + 0.0001T + 0.2303$	0.9997
Zircon, $\rho_t = 3{,}580$ kg/m^3	$0 \leq T°C \leq 1{,}400$	$\lambda_t(T) = 8 \cdot 10^{-10}T^3 - 6 \cdot 10^{-8}T^2 + 6 \cdot 10^{-6}T + 2.7991$	0.9999
Zircon, $\rho_t = 1{,}500$ kg/m^3	$0 \leq T°C \leq 1{,}400$	$\lambda_t(T) = 2 \cdot 10^{-7}T^2 - 8 \cdot 10^{-5}T + 0.6342$	0.9901

From: [23].

References

[1] Arai, M., et al., "Elevated Temperature Dielectric Property Measurements: Results of a Parallel Measurement Program," *Ceramic Transactions, Microwaves: Theory and Application in Material Processing II*, Vol. 36, 1993, pp. 539–546.

[2] Braunstein, J., K. Connor, and S. Salon, et al., "Investigation of Microwave Heating with Time Varying Materials Properties in Ceramic Processing," *IEEE Transactions on Magnetics*, Vol. 35, 1999, pp. 1813–1816.

[3] Hamlyn, M. G., A. L. Bowden, and N. G. Evans, "Measurement and Use of High Temperature Dielectric Properties in Ceramic Processing," *Microwave and High Frequency Heating, International Conference.* Cambridge, United Kingdom, 1995, p. N1.1.

[4] Xi, W., and W. Tinga, "Error Analysis and Permittivity Measurements with Re-Entrant High-Temperature Dielectrometer," *Journal of Microwave Power and Electromagnetic Energy*, Vol. 28, No. 2, 1993, pp. 104–112.

[5] Xi, W., and W. R. A. Tinga, "A High Temperature Microwave Dielectrometer," *Proceedings of the Symposium on Microwaves: Theory and Application in Material Processing*, Cincinnati, OH, 1991, pp. 215–224.

[6] FEMLAB V. 2.3, Comsol (www.comsol.com), *Electromagnetic Module*, Model Library, 2003.

[7]   Baeraky, T. A., "Microwave Measurements of Dielectric Properties of Zinc Oxide at High Temperatures," *Egyptian Journal of Solids*, Vol. 30, No. 11, 2007, pp. 13–18.

[8]   Antsiferov, A. A., et al., "Set Up for Control of Parameters of Dielectrics in Microwave Range," *Elektronnaya tekhnika. Electronika SVCH*, Vol. 9, 1963, pp. 138–142 (in Russian).

[9]   Evans, N. G., and M. G. Hamlyn, "Alkali Metal and Alkali Earth Carbonates at Microwave Frequencies, I: Dielectric Properties," *Journal of Microwave Power and Electromagnetic Energy*, Vol. 33, No. 1, 1998, pp. 24–26.

[10] Evans, N. G., and M. G. Hamlyn, "Alkali Metal and Alkali Earth Carbonates at Microwave Frequencies, II: Microwave Heating of Barium Carbonate," *Journal of Microwave Power and Electromagnetic Energy*, Vol. 33, No. 1, 1998, pp. 27–30.

[11] Jerby, E., O. Aktushev, and V. Dikhtyar, "Theoretical Analysis of the Microwave Drill Near Field Localized Effect," *Journal of Applied Physics*, Vol. 97, No. 3, 2004, p. 034909.

[12] Chudnovskiy, A. F., *Thermo Physical Characteristics of Disperse Materials*, Moscow: Izdatelstvo Fizikomatematicheskoy Literaturi, 1962 (in Russian).

[13] Platunov, E. S., S. E. Buravoy, and V. V. Kurepin, et al., *Thermophysical Measurements and Instruments*, Leningrad: Mashinostroenie Issue, 1986 (in Russian).

[14] Donaldson, K. Y., B. Trandell, and Y. Lu, et al., "Effect of Felamination on the Transverse Thermal Conductivity of SiC Fiber Reinforced SiC-Matrix Composite," *Journal of American Ceramics Society*, 1998, Vol. 81, No. 6, pp. 1583–1588.

[15] Yamada, R., T. Taguchi, and J. Nakano, et al., "Thermal Conductivity of CVI PIP SIC/SIC Composites," *Proceedings of the 23rd International Conference on Composites, Advanced Ceramics, Materials and Structures*, 1999, pp. 273–280.

[16] Prietzel, S., and A. Lipp, "Boron Nitride Powder: Thermal and Electrical Application," *High Tech. Ceramics. Proceedings of the 6th International Workshop on Modern Ceramics Technologies*, Italy, 1986, pp. 2337–2341.

[17] Inomata, Y., "Thermal Conductivity of Si_3N_4, AlN and Si-Al-O-N Ceramics," *Energy and Ceramics: Proceedings of the 4th International Workshop on Modern Ceramics Technologies*, Italy, 1979, pp. 706–713.

[18] Mazurin, O. V., "Silica Glass and Binary Silicate Glasses," *Handbook of Glass Data*, New York: Elsevier Science, 1983.

[19] Shackelford, J. F. and W. Alexander (eds.), *CRC Materials Science and Engineering Handbook, 3rd Edition*, New York: CRC Press, 2001.

[20] Sorai, M. (ed.), *Comprehensive Handbook of Calorimetry and Thermal Analysis*, Hoboken, NJ: John Wiley and Sons, Inc., 2005.

[21] Lax, D. (ed.), *Taschenbuch fur Chemiker und Physiker*, Band 1. Berlin: Springer-Verlag, 1967 (in German).

[22] Grigoriev, I. S. and E. Z. Meylihov (eds.), *Physical Constants: Handbook*, Moscow: Energoatomizdat, 1991 (in Russian).

[23] Hirschberg, H. G., *Handbook Verfahrens-technik und Anlagenbau*, Berlin: Springer, 1999 (in German).

8

Soils and Minerals

Moisture-dependent dielectric constants of many soils (sandy, high-clay, loamy, etc.) over RF and microwave frequency ranges have been published by several authors [1–9]. Most of these studies have been carried out in the temperature range ($5 \leq T°C \leq 25$) for fixed values of soil density. As an example in Tables 8.1–8.8, the complex dielectric permittivity of several types of soils is listed for different moisture content at microwave frequencies [3–7].

Table 8.1
Dielectric Properties of Sand and Clay
at Room Temperature

W(%)	f (GHz)	Sand ε'	ε''	Clay ε'	ε''
4	1	3.33	0.13	3.05	0.78
	3	3.05	0.26	2.88	0.73
12	1	11.6	0.78	12.5	4.3
	3	10.8	1.04	11.38	2.3
20	1	20.3	1.17	22.2	8.78
	3	19.4	1.96	19.2	4.44

From: [4].

Table 8.2
Complex Dielectric Permittivity of Dry and Moist Soils
at 2.45 GHz and Room Temperature

Moisture content	W = 0%	W = 10%	W = 20%	W = 30%
Dielectric permittivity	2.3	4.07	6.4	9.6
Loss factor	0.05	0.4	0.7	1.25

From: [5].

Table 8.3
Dielectric Properties of Soil Moist Salts at 1.11 GHz and $T = 24°C$

Salt	Density (g/cm^3)	W*	Expression
Na$_2$CO$_3$	1.1 ÷ 1.3	0 ÷ 0.61	$\varepsilon' = ((1.56 \pm 0.04) + (0.69 \pm 0.09)\ W*)^2$
			$\varepsilon'' = ((0.01 \pm 0.005) + (0.03 \pm 0.02)\ W*)^2$
Na$_2$SO$_4$	1.3 ÷ 1.6	0 ÷ 0.57	$\varepsilon' = ((1.59 \pm 0.03) + (0.5 \pm 0.01)\ W*)^2$
			$\varepsilon'' = ((0.02 \pm 0.005) + (0.05 \pm 0.02)\ W*)^2$
MgSO$_4$	1.0 ÷ 1.3	0 ÷ 0.59	$\varepsilon' = ((1.63 \pm 0.02) + (0.3 \pm 0.09)\ W*)^2$
			$\varepsilon'' = ((0.03 \pm 0.01) + (0.01 \pm 0.02)\ W*)^2$

From: [6].
W* = $(M_w - M_d)/V$, where M_w = moist sample mass; M_d = dry sample mass; V = sample volume.

Table 8.4
Dielectric Properties of Moist Minerals at 1.11 GHz and $T = 24°C$

Mineral	W*	Expression
Quartz	$0.005 \leq W* \leq 0.5$	$\varepsilon' = ((1.54 \pm 0.05) + (9.2 \pm 0.4)\ W*)^2$
		$\varepsilon'' = ((0.01 \pm 0.005) + (0.79 \pm 0.064)\ W*)^2$
Calcite	$0.02 \leq W* \leq 0.5$	$\varepsilon' = ((1.73 \pm 0.06) + (8.1 \pm 0.2)\ W*)^2$
		$\varepsilon'' = ((0.01 \pm 0.005) + (0.37 \pm 0.05)\ W*)^2$
Kaolinite	$0.005 \leq W* \leq 0.12$	$\varepsilon' = ((1.9 \pm 0.1) + (3.3 \pm 0.9)\ W*)^2$
		$\varepsilon'' = ((0.2 \pm 0.01) + (0.52 \pm 0.09)\ W*)^2$
	$0.12 \leq W* \leq 0.6$	$\varepsilon' = ((1.1 \pm 0.1) + (8.0 \pm 0.2)\ W*)^2$
		$\varepsilon'' = ((0.18 \pm 0.02) + (0.48 \pm 0.06)\ W*)^2$
Marcasite	$0.005 \leq W* \leq 0.18$	$\varepsilon' = ((1.1 \pm 0.03) + (5.26 \pm 0.07)\ W*)^2$
		$\varepsilon'' = ((0.04 \pm 0.02) + (0.25 \pm 0.05)\ W*)^2$
	$0.18 \leq W* \leq 0.6$	$\varepsilon' = ((0.74 \pm 0.08) + (7.2 \pm 0.3)\ W*)^2$
		$\varepsilon'' = ((0.04 \pm 0.02) + (0.25 \pm 0.05)\ W*)^2$
Montmorillonite	$0.005 \leq W* \leq 0.18$	$\varepsilon' = ((1.63 \pm 0.03) + (8.6 \pm 0.4)\ W*)^2$
		$\varepsilon'' = ((0.01 \pm 0.005) + (6.4 \pm 0.4)\ W*)^2$
	$0.18 \leq W* \leq 0.6$	$\varepsilon' = ((1.5 \pm 0.1) + (9.9 \pm 0.4)\ W*)^2$
		$\varepsilon'' = ((0.02 \pm 0.01) + (6.4 \pm 0.2)\ W*)^2$

From: [6].

Table 8.5
Dielectric Properties of Moist Salty Sands with Salinity 5% in Moisture Range
$0 \leq W^* \leq 0.3$ at 1.11 GHz and $T = 24°C$

Soil	Density of Dry Soil (g/cm³)	Expression
Sand	1.51 ± 0.07	$\varepsilon' = ((1.71 \pm 0.02) + (8.1 \pm 0.1)\ W^*)^2$
		$\varepsilon'' = ((0.03 \pm 0.001) + (0.5 \pm 0.01)\ W^*)^2$
Sand + NaCl 5% Salt content	1.63 ± 0.03	$\varepsilon' = ((1.56 \pm 0.05) + (16 \pm 0.8)\ W^*)^2$
		$\varepsilon'' = ((0.001 \pm 0.0001) + (19 \pm 1)\ W^*)^2$

From: [7].

Table 8.6
Dielectric Properties of Moist Klinoptiolite at 1.11 GHz and $T = 24°C$

Soil	Moisture	Expression
Klinoptiolite	$W^* \leq 0.05$	$\varepsilon' = ((1.62 \pm 0.01) + (4.6 \pm 0.5)\ W^*)^2$
		$\varepsilon'' = ((0.01 \pm 0.001) + (1.5 \pm 0.3)\ W^*)^2$
	$0.05 \leq W^* \leq 0.12$	$\varepsilon' = ((1.71 \pm 0.03) + (2.8 \pm 0.3)\ W^*)^2$
		$\varepsilon'' = ((0.01 \pm 0.001) + (1.3 \pm 0.2)\ W^*)^2$
	$0.12 \leq W^* \leq 0.25$	$\varepsilon' = ((0.98 \pm 0.04) + (9.0 \pm 0.2)\ W^*)^2$
		$\varepsilon'' = ((0.1 \pm 0.02) + (0.52 \pm 0.1)\ W^*)^2$
Klinoptiolite + NaCl (5%)	$0 \leq W^* \leq 0.23$	$\varepsilon' = ((1.62 \pm 0.01) + (2 \pm 0.4)\ W^* + (4.2 \pm 0.3)\ W^{*2})^2$
		$\varepsilon'' = ((0.02 \pm 0.005) - (0.6 \pm 0.1)\ W^* + (50 \pm 3)W^{*2})^2$

From: [7].

Table 8.7
Dielectric Properties of Some Moist Soils ($0.05 \leq W \leq 0.5$) at 5 GHz and Room Temperature

Soil Texture (%)			Expression	R^2
Sand	Silt	Clay		
88	7.3	4.7	$\varepsilon'(W) = -219.88W^3 + 135.22W^2 + 17.569W + 2.8875$	0.9994
			$\varepsilon''(W) = -6.8833W^2 + 9.238W - 0.2731$	0.9925
56	26.7	17.3	$\varepsilon'(W) = -386.57W^3 + 297.01W^2 - 18.996W + 4.3058$	0.9984
			$\varepsilon''(W) = 5.9384W - 0.0983$	0.9952
19.3	46	34.7	$\varepsilon'(W) = -82.542W^3 + 108.94W^2 + 4.8748W + 2.1982$	0.9955
			$\varepsilon''(W) = -57.171W^3 + 58.452W^2 - 4.4717W + 0.3899$	0.9980
2	37	61	$\varepsilon'(W) = -149.6W^3 + 158.7W^2 - 2.3992W + 2.1922$	0.9967
			$\varepsilon''(W) = 6.3315W - 0.1221$	0.9969

From: [3].
W = volumetric water content (cm³/cm³).

Table 8.8
Dielectric Properties of Some Moist Soils at 1.412 GHz and Room Temperature

Soil Texture (%)				
Sand	Silt	Clay	Expression	R^2
100	0	0	$\varepsilon'(W) = 122.31W^2 + 20.451W + 3.0646$	0.9998
$0.05 \leq W \leq 0.3$			$\varepsilon''(W) = 4.8451W - 0.1467$	0.9944
90	7	3	$\varepsilon'(W) = 180.39W^2 + 6.1294W + 3.4218$	0.9989
$0.05 \leq W \leq 0.25$			$\varepsilon''(W) = -5.5429W^2 + 4.9389W - 0.1116$	0.9934
22	70	8	$\varepsilon'(W) = -383.33W^3 + 309.96W^2 - 17.191W + 4.2391$	0.9996
$0.05 \leq W \leq 0.35$			$\varepsilon''(W) = 11.399W^2 + 2.4621W + 0.2905$	0.9993
58	28	14	$\varepsilon'(W) = 73.286W^2 + 35.407W + 0.17$	0.9993
$0.05 \leq W \leq 0.35$			$\varepsilon''(W) = 15.757W^2 + 1.7779W + 0.2956$	0.9991
6	54	40	$\varepsilon'(W) = 147.73W^2 - 20.059W + 5.7151$	0.9989
$0.05 \leq W \leq 0.4$			$\varepsilon''(W) = 42.077W^2 - 5.229W + 1.0798$	0.9978

From: [3].

At 2,450 MHz, all dry sandy soils may be classified into three groups according to the value of loss factor: sands with low losses ($\varepsilon'' \leq 0.0075$), sands with average losses ($0.011 \leq \varepsilon'' \leq 0.031$), and sands with high losses ($0.05 \leq \varepsilon'' \leq 0.1$). Dry clay soils are characterized by $3.91 \leq \varepsilon' \leq 4.21$ and $0.14 \leq \varepsilon'' \leq 0.32$. The dielectric properties of loamy soils with moisture content are $2 \leq MC, \% \leq 50$ and $3.49 \leq \varepsilon' \leq 32.53$ and $0.22 \leq \varepsilon'' \leq 1.3$.

EM fields are used today in the mining industry, geology, road-building technologies, and so forth. Determining the dielectric properties of minerals allows more accurate mathematical modeling of the physical processes of EM wave interaction with lossy media. The open-end coaxial line method has been utilized in [8] for measuring $\varepsilon'(f, T)-j\varepsilon''(f, T)$ of over 60 minerals in temperature range between 25°C and 325°C and frequency range $300 \leq f$, MHz $\leq 3,000$. Some of this experimental data at 2,450 GHz is given in Table 8.9. Most of the minerals exhibit increasing value of ε' and ε'' versus temperature, except for minerals presented in Table 8.9, which have complicated dependencies $\varepsilon'(T)$ and $\varepsilon''(T)$ [8]. The temperature-dependent CDP values of dry and moist montmorillonite at 1.11 GHz and sand at 1.1 GHz are given in Tables 8.10 and 8.11, respectively.

Table 8.9

Dielectric Properties of Minerals

Mineral	f (GHz)	$T°C$	25	100	200	300
Chalcocite	0.915	ε'	17.14	17.64	15.44	11.95
		ε''	1.132	1.28	1.13	1.135
	2.45	ε'	16.72	17.14	15.46	11.82
		ε''	1.64	1.76	1.52	1.12
Chalcopyrite	0.915	ε'	12.9	13.64	13.15	9.4
		ε''	10.89	14.19	17.97	0.81
	2.45	ε'	11.01	11.4	11.2	9.0
		ε''	8.0	10.2	12.5	1.4

From: [8].

Table 8.10

Dielectric Properties of Montmorillonite

Humidity	Temperature	Expression
0.04	$20 < T°C \leq 110$	$\varepsilon'(T) = (4.01 \pm 0.03) - (0.007 \pm 0.001)\,T + (1.17 \pm 0.06) \cdot 10^{-4} T^2$
		$\varepsilon''(T) = (1.1 \pm 0.03) - (0.007 \pm 0.001)\,T + (1.56 \pm 0.08) \cdot 10^{-4} T^2$
	$114 \leq T°C < 140$	$\varepsilon'(T) = (3.43 \pm 0.09) + (0.014 \pm 0.001)\,T$
		$\varepsilon''(T) = (0.4 \pm 0.1) + (0.018 \pm 0.001)\,T$
0.45	$10 \leq T°C \leq 80$	$\varepsilon'(T) = (21.9 \pm 0.14) - (0.019 \pm 0.03)\,T$
		$\varepsilon''(T) = (6.9 \pm 0.2) + (0.027 \pm 0.003)\,T$

From: [9].

Table 8.11

Dielectric Properties of Sand at 1.1 GHz in Temperature Range $0 < T°C \leq 40$

W%	Dielectric Permittivity	Loss Factor	R^2
5	$\varepsilon'(T) = 3.08$	$\varepsilon''(T) = 0.155 - 0.002T$	0.9999
10	$\varepsilon'(T) = 6.7 - 0.0088T$	$\varepsilon''(T) = 0.0001T^2 - 0.0098T + 0.3444$	0.9984
15	$\varepsilon'(T) = 9.04 - 0.0193T$	$\varepsilon''(T) = 0.0002T^2 - 0.0185T + 0.5773$	0.9959
20	$\varepsilon'(T) = 11.9 - 0.0283T$	$\varepsilon''(T) = 0.0003T^2 - 0.0252T + 0.8255$	0.9973
25	$\varepsilon'(T) = 14.8 - 0.0335T$	$\varepsilon''(T) = 0.0004T^2 - 0.0352T + 1.1512$	0.9984
30	$\varepsilon'(T) = 18.08 - 0.0432T$	$\varepsilon''(T) = 0.0005T^2 - 0.0469T + 1.5184$	0.9979

From: [10].

The thermal parameters of soils and minerals have been studied in many publications (see for example [11–17]). Some of this data is presented in Tables 8.12–8.17. One substance, ice, has been included in this section as the exclusion (Table 8.13): the CDP of ice at 2.45 GHz: $\varepsilon' = 3.2$; $\varepsilon'' = 0.003$ [18].

The heat capacity $(J/g \cdot K)$ of several rock soils—magnesium silicate, calcium feldspar, diabase, diorite, granite, basalt, silica glass, and quartz—in temperature range $100 \leq T°K \leq 500$ is described by one curve [13]:

$$C_t(T) = 0.482 \ln(T) - 1.9806, \quad R^2 = 0.9952 \tag{8.1}$$

Table 8.12
Thermal Properties of Some Rocks at Room Temperature

Rock	ρ_t (kg/m³)	C_t (kJ/kg·K)	λ_t (W/m·K)
Brown coal	1,210	1.13	0.253
Anthracite	1,440	0.946	0.328
Coal schist	1,765	1.0216	0.835
Clay schist	2,433	0.992	0.9315
Albitofir	2,596	0.979	1.983
Granite	2,722	0.917	2.214
Dolomite	2,675	0.929	1.729
Dense sandstone	2,630	0.963	2.004
Dense chalk	1,780	0.917	0.688
Dense limestone	2,478	0.8876	0.984

From: [11].

Table 8.13
Thermal Conductivity of Some Soils with Density $1,570 \leq \rho_t$, kg/m³ $\leq 1,760$

$T°C$	100	200	400	600	800	1,000	1,200	1,300	1,400
Limestone	0.79	0.77	0.74	0.71	0.68	0.66	0.64	0.63	0.64
Granite detrital	0.9	0.88	0.83	0.8	0.79	0.79	0.81	0.84	0.91
Sea sand	0.29	0.31	0.34	0.37	0.39	0.43	0.52	0.6	0.71
Magnesium	0.66	0.66	0.65	0.65	0.65	0.66	0.69	0.71	0.76
Podzol	0.51	0.52	0.52	0.52	0.53	0.58	0.7	0.79	0.91
Coastal plains	0.27	0.28	0.3	0.33	0.35	0.39	0.43	0.46	0.51
Laterite	0.34	0.31	0.27	0.23	0.2	0.2	0.26	0.32	0.42

From: [12].

Table 8.14
Thermal Properties: $\lambda_t (W/m \cdot K)$, C_t $(J/kg \cdot K)$, ρ_t (kg/m^3) of Ice

Parameter	R^2	Temperature
$\lambda_t(T) = 488.19T^{-1} + 0.4685$ [12]	0.9999	$73 \leq T°K \leq 273$
$C_t(T) = 7 \cdot 10^{-5}T^3 - 0.0365T^2 + 12.481T - 119.54$	0.9998	$20 \leq T°K \leq 273$
$\lambda_t(T) = 4 \cdot 10^{-5}T^2 - 0.0292T + 7.4651$ [13]	0.9993	$173 \leq T°K \leq 273$
$C_t(T) = 1500\ln(T) - 6354.5$	0.9971	
$\rho_t(T) = 940.37 - 0.0841T$	0.9741	

Table 8.15
Thermal Conductivity $(W/m \cdot K)$ of Different Soils with Saturation $10 \div 100\%$

Soil	Parameter	Temperature
Ottawa dry sand with density 1,760 kg/m^3	$\lambda_t(T) = 6 \cdot 10^{-8}T^2 + 0.0004T + 0.3643$, $R^2 = 0.9998$	$100 \leq T°C \leq 1400$
Unfrozen silt and clay, $2 \leq W\% \leq 50$	$0.2 \leq \lambda_t \leq 1.8$ $800 \leq \rho_t$, $kg/m^3 \leq 1,800$	$20 \leq T°C \leq 100$
Frozen silt and clay	$0.5 \leq \lambda_t \leq 2.2$ $800 \leq \rho_t$, $kg/m^3 \leq 2,000$	$173 \leq T°K \leq 273$
Unfrozen sandy soils, $10 \leq W\% \leq 100$	$0.5 \leq \lambda_t \leq 3$ $1,000 \leq \rho_t$, $kg/m^3 \leq 2,000$	$20 \leq T°C \leq 100$
Frozen sandy soils	$1 \leq \lambda_t \leq 4$ $1,000 \leq \rho_t$, $kg/m^3 \leq 2,000$	$173 \leq T°K \leq 273$
Dry sand	$\lambda_t(T) = 0.00049T + 0.24$	$283 \leq T°K \leq 373$

From: [13].

Table 8.16
Thermal Conductivity (W/m·K) of soils

Soil	Temperature Range	Thermal Conductivity
Fine-grained sand	$261 \leq T \, °K \leq 272$	2.475
	$273 \leq T \, °K \leq 283$	1.95
Wood loam	$261 \leq T \, °K \leq 271$	1.75
	$273 \leq T \, °K \leq 283$	1.2
Clay	$261 \leq T \, °K \leq 271$	1.3 ÷ 1.5
	$273 \leq T \, °K \leq 283$	1.2
Peat	$261 \leq T \, °K \leq 271$	1.2 ÷ 1.4
	$273 \leq T \, °K \leq 283$	0.9
Hard loamy	$261 \leq T \, °K \leq 271$	2 ÷ 2.15
	$273 \leq T \, °K \leq 285$	1.4 ÷ 1.5
Hard quartz loamy	$259 \leq T \, °K \leq 271$	2.5 ÷ 2.6
	$273 \leq T \, °K \leq 287$	1.6 ÷ 1.8
Koalinite clay	$259 \leq T \, °K \leq 271$	0.9 ÷ 1.15
	$273 \leq T \, °K \leq 287$	0.6

From: [16].

Table 8.17
Heat Capacity (J/g·K) of Minerals

Mineral	Temperature (°C)	C_t	Mineral	Temperature (°C)	C_t
Orthoclase	$-20 \leq T \leq 100$	0.776	Lava	$23 \leq T \leq 100$	0.858
Albite	$0 \leq T \leq 900$	1.072	Magnesite	$T = 18$	0.889
Andalusite	$0 \leq T \leq 100$	0.705	Magnetite	$0 \leq T \leq 200$	0.746
Anorthite	$0 \leq T \leq 900$	1.038	Pyrrhotite	$0 \leq T \leq 100$	0.626
Asbestos	$0 \leq T \leq 34$	0.783	Magnetic silicate	$0 \leq T \leq 900$	1.114
Basalt	$10 \leq T \leq 100$	0.585	Manganite	$20 \leq T \leq 52$	0.737
Beryllium	$15 \leq T \leq 99$	0.828	Paragonite	$20 \leq T \leq 98$	0.873
Braunite	$15 \leq T \leq 99$	0.678	Olivine	$21 \leq T \leq 51$	0.791
Calcite	$T = 21$	0.813	Orthoclase	$15 \leq T \leq 99$	0.786
Chalcedony	$0 \leq T \leq 1000$	1.102	Scheelite	$19 \leq T \leq 50$	0.405
Dioptase	$19 \leq T \leq 50$	0.762	Serpentine	$8 \leq T \leq 98$	1.067
Diopside	$0 \leq T \leq 100$	0.805	Argentite	$15 \leq T \leq 100$	0.309
Gneis	$17 \leq T \leq 99$	0.82	Spinel	$15 \leq T \leq 46$	0.812
Goethite	$5 \leq T \leq 93$	0.854	Spodumene	$20 \leq T \leq 100$	0.905
Granite	$16 \leq T \leq 100$	0.829	Rock salt	$13 \leq T \leq 45$	0.917
Graphite	$20 \leq T \leq 200$	0.984	Talc	$20 \leq T \leq 98$	0.876
Hornblende	$20 \leq T \leq 98$	0.817	Topaz	$6 \leq T \leq 100$	0.858
Ilmenite	$17 \leq T \leq 47$	0.741	Wollastonite	$0 \leq T \leq 900$	0.981
Calcite	$0 \leq T \leq 100$	0.839	Sphalerite	$0 \leq T \leq 100$	0.48
Cobaltite	$15 \leq T \leq 100$	0.406	Zircon	$21 \leq T \leq 51$	0.552

From: [15].

Table 8.18
Heat Capacity (kJ/kg·K) of Some Minerals and Soils

Soil		C_t	Soil		C_t
Andalusite	$273 \leq T \, °K \leq 373$	0.7	Dolomite	$293 \leq T \, °K \leq 372$	0.93
Apatite	$288 \leq T \, °K \leq 372$	0.79	Kaoline	$T = 273 \, °K$	1
Basalt	$T = 273 \, °K$	0.85		$T = 673 \, °K$	1.35
	$T = 1473 \, °K$	1.49	Lava	$296 \leq T \, °K \leq 373$	0.84
Gneiss	$T = 273 \, °K$	0.74		$304 \leq T \, °K \leq 1049$	1.09
	$T = 473 \, °K$	1.02	Spinel	$282 \leq T \, °K \leq 371$	0.81
Granite	$T = 273 \, °K$	0.65	Clay	$T = 273 \, °K$	0.75
	$T = 1073 \, °K$	1.3		$T = 1073 \, °K$	1.51
Natural graphite	$T = 300 \, °K$	$0.95 \div 1.05$	Limestone	$273 \leq T \, °K \leq 373$	0.92

From: [15].

References

[1]   Hoekstra, P., and A. Delaney, "Dielectric Properties of Soils at UHF and Micro-wave Frequencies," *Journal of Geophysics Research*, Vol. 79, 1974, pp. 1699–1708.

[2]   Hipp, J. E., "Soil Electromagnetic Parameters as a Function of Frequency, Soil Density and Soil Moisture," *Proceedings of IEEE*, Vol. 62, 1974, pp. 98–103.

[3]   Wang, J. R., and T. J. Schmugge, "An Empirical Model for the Complex Dielectric Permittivity of Soils as a Function of Water Content," *IEEE Transactions on Geosciences Remote Sensing*, Vol. GE-18, 1980, pp. 288–295.

[4]   Shutko, A. M., *Microwave Radio Sensing of Water Surfaces and Soils*, Moscow: Nau-ka, 1986 (in Russian).

[5]   Pauli, M., T. Kayser, and W. Wiesbeck, "A Versalite Measurement System for the Determination of Dielectric Parameters of Various Materials," *Measurement Science and Technology*, Vol. 18, 2007, pp. 1046–1053.

[6]   Romanov, A. N., "Influence of Mineralogical Composition on Dielectric Properties of Dispersive Mixtures in the Microwave Range," *Journal of Communication Technology and Electronics*, Vol. 48, No. 5, 2003, pp. 487–493.

[7]   Romanov, A. N., "A Model for the Humidity Dependence of the Microwave Complex Permittivity of Saline Soils," *Journal of Communication Technology and Electronics*, Vol. 51, No. 12, 2006, pp. 1379–1386.

[8]   Salsman, J. B., "Measurement of Dielectric Properties in the Frequency Range of 300 MHz to 3 GHz as a Function of Temperature and Density," *Proceedings of the Symposium on Microwaves: Theory and Application in Material Processing*, Cincinnati, OH, 1991, pp. 215–224.

[9]   Romanov, A. N., "Effect of the Thermodynamic Temperature on the Dielectric Characteristics of Mineral and Bound Water in the Microwave Band," *Journal of Communication Technology and Electronics*, Vol. 49, No. 1, 2004, pp. 83–87.

[10]   Jaganathan, A. P., and E. N. Allouche, "Temperature Dependence of Dielectric Properties of Moist Soils," *Canadian Geotechnical Journal*, Vol. 45, 2008, pp. 888–894.

[11]   Chudnovskiy, A. F., *Thermo Physical Characteristics of Disperse Materials*, Moscow: Izdatelstvo Fizikomatematicheskoy literaturi, 1962 (in Russian).

[12]   Flynn, D. R., and T. W. Watson, "High-Temperature Thermal Conductivity of Soils," *Proceedings of the 8th International Conference on Thermal Conductivity*, Purdue University, West Lafayette, IN, October 7–10, 1968, pp. 913–939.

[13]   Farouki, O. T., *Thermal Properties of Soils. Series on Rock and Soil Mechanics*, Trans Tech Publications, 1986.

[14]   www.engineeringtoolbox.com.

[15]   Lax, D. (ed.), *Taschenbuch fur chemiker und physiker, Band 1*, Berlin: Springer-Verlag. 1967 (in German).

[16] Yershov, E. D. (ed.), *Thermophysical Properties of Rocks*, Moscow: Moscow State University Publishing, 1984 (in Russian).

[17] Grigoriev, I. S., and E. Z. Meylihov (eds.), *Physical Constants: Handbook*, Moscow: Energoatomizdat, 1991 (in Russian).

[18] Yang, X. H., and J. Tang (eds.), *Advances in Bioprocessing Engineering*, Vol. 1, London: World Scientific, 2002.

9

Pure and Composite Chemical Substances

Today microwave radiation is widely utilized in various technologies for accelerating many chemical reactions of organic and inorganic synthesis, dehydration, digestion, curing, vulcanization, and so forth. Some of these reactions can be exothermic. But irrespective of the type of chemical reaction, the physical properties of reactants usually strongly depend on temperature. Viscosity is a very important parameter that influences temperature distribution when liquid chemical substances are heated with microwave radiation.

Water is a very well-known substance. The temperature-dependent dielectric properties of water are widely available in the literature, but in many cases it is difficult to find this data exactly at ISM frequencies, or it can only be found in limited temperature intervals.

The mechanisms of EM wave interaction with water and similar substances such as dipole polarization, heat release, and ionic conduction are well described in [1, 2]. The complex dielectric permittivity of pure water is determined by the Debye model [1]:

$$\dot{\varepsilon} = \varepsilon' - j\varepsilon'' = \varepsilon_\infty + \frac{\varepsilon_s - \varepsilon_\infty}{1 + \omega^2\tau^2} - j\frac{\left(\varepsilon_s - \varepsilon_\infty\right)\omega\tau}{1 + \omega^2\tau^2} \tag{9.1}$$

where ε_∞ is the infinite or high-frequency relative permittivity, ε_s is the static or zero-frequency relative permittivity, ω is the angular frequency, and τ is the relaxation time; $j = \sqrt{-1}$.

Water, which contains salts and other admixtures, also has ion conductivity, and (9.1) should be completed by one more term:

$$\dot{\varepsilon} = \varepsilon_\infty + \frac{\varepsilon_s - \varepsilon_\infty}{1 + \omega^2 \tau^2} - j\frac{(\varepsilon_s - \varepsilon_\infty)\omega\tau}{1 + \omega^2\tau^2} - j\frac{\sigma}{\omega\varepsilon_0} \qquad (9.2)$$

where σ is the electrical conductivity.

The CDP of water as a function of temperature (T°C) was studied by many authors, including [2–10]. Some measured $\varepsilon'(T)$ and $\varepsilon''(T)$ [2, 4, 6, 9] and others [3, 5] developed mathematical models based on (9.1) and (9.2). A comparison of several theoretical and experimental results of such studies at $915 \div 930$ MHz and $2.45 \div 3$ GHz is given in Tables 9.1 to 9.4.

Table 9.1
Complex Dielectric Permittivity of Tap Water at 915 MHz

T°C	20	30	40	50	60
ε'	79.1 ± 2.4	74.4 ± 1.4	69.6 ± 1.4	65.4 ± 1.2	62.2 ± 0.7
ε''	6.6 ± 1.4	5.6 ± 1	5 ± 0.8	4.5 ± 0.9	4.1 ± 0.8

From: [10].

Table 9.2
Complex Dielectric Permittivity of Tap Water at 915 MHz

Salinity (‰)	T°C	2	5	10	15	20	25	30	35	40
$0 \div 0.05$	ε'	86.0	85.3	83.7	81.9	79.9	78.0	76.4	75.3	74.5
$0 \div 0.03$	ε''	7.75	6.95	5.79	4.85	4.10	3.51	3.07	2.73	2.48
0.05 ‰		7.77	6.97	5.82	4.88	4.13	3.55	3.11	2.77	2.53
0.05 ‰		7.79	7.00	5.85	4.91	4.16	3.58	3.15	2.82	2.57

From: [3].

Table 9.3
Dielectric Properties of Tap Water at Microwave Frequencies

$T°C$	−9.9	−7.3	−5.2	15	19.75	39.85	40.75	60.35	80.1	90.1
f, MHz	918	919	920	921	923	929	925	931	933	933
ε'	91.61	90.02	88.8	85.4	80.16	73.11	72.8	66.717	60.865	58.04
ε''	13.15	11.83	10.86	7.01	4.156	2.318	2.254	1.449	0.95	0.736

From: [6].

Table 9.4
Dielectric Properties of Water at Different Microwave Frequencies

f (GHz)	Temperature	Expression	R^2
0.915 [9]	$40 \leq T°C \leq 120$	$\varepsilon'(T) = -16.308\ln(T) + 133.75$	0.9962
		$\varepsilon''(T) = -3.68 \cdot 10^{-6}T^3 + 1.03 \cdot 10^{-3}T^2 - 0.0947T + 5.6$	0.9891
3 [2]	$5 \leq T°C \leq 85$	$\varepsilon'(T) = -0.0017T^2 - 0.1484T + 80.951$	0.9990
		$\varepsilon''(T) = 24.342\exp(-0.0248T)$	0.9951
3 [7]	$0 \leq T°C \leq 60$	$\varepsilon'(T) = -0.0023T^2 - 0.0989T + 80.395$	0.9988
		$\varepsilon''(T) = -7 \cdot 10^{-5}T^3 + 0.011T^2 - 0.7414T + 23.965$	0.9999
1.744 [7]		$\varepsilon'(T) = -0.3125T + 85.418$	0.9999
		$\varepsilon''(T) = -8 \cdot 10^{-5}T^3 + 0.0111T^2 - 0.6218T + 16.412$	0.9992
3.254 [7]		$\varepsilon'(T) = -0.003T^2 - 0.529T + 80.062$	0.9998
		$\varepsilon''(T) = -9 \cdot 10^{-5}T^3 + 0.0137T^2 - 0.8701T + 26.588$	0.9994
2.45 [4]	$20 \leq T°C \leq 70$	$\varepsilon'(T) = 8 \cdot 10^{-5}T^3 - 0.0125T^2 + 0.3072T + 75.756$	0.9825
		$\varepsilon''(T) = 0.0018T^2 - 0.2826T + 13.809$	0.9906

The complex dielectric permittivity of distilled water at 2.8 GHz in temperature range $5 \leq T°C \leq 60$ [11] is:

$$\varepsilon'(T) = -7.2883 \cdot 10^{-5}T^3 + 0.0057366T^2 - 0.42656T + 84.4515 \tag{9.3}$$

$$\varepsilon''(T) = -4.2471 \cdot 10^{-5}T^3 + 0.0097473T^2 - 0.71507T + 22.9926 \tag{9.4}$$

The complex dielectric permittivity of fresh water at 2.45 GHz in temperature range $20 \leq T°C \leq 100$ [12] is:

$$\varepsilon'(T) = -4.6 \cdot 10^{-6}T^3 + 0.00131T^2 - 0.414T + 88.15 \tag{9.5}$$

$$\varepsilon''(T) = -5 \cdot 10^{-5}T^3 + 0.0103T^2 - 0.8064T + 26.675 \tag{9.6}$$

Table 9.5
Thermal Properties of Water in Temperature Range $10 \leq T°C \leq 100$

Parameter	Expression	Units	R^2
Density	$\rho_t(T) = -0.0043T^2 - 0.0109T + 999.79$	kg/m^3	0.9979
Heat capacity	$C_t(T) = -10^{-4}T^3 + 0.0309T^2 - 2.0378T + 4210.8$	J/kg·K	0.9988
Thermal conductivity	$\lambda_t(T) = -10^{-5}T^2 + 0.0024T + 0.5565$	W/m·K	0.9994
Kinematical viscosity	$\nu_t(T)\ 10^6 = 2.3342 - 0.4493\ln(T)$	m^2/s	0.9974
Dynamic viscosity	$\mu_t(T)\ 10^6 = 2348 - 454.69\ln(T)$	Pa·s	0.9976
Prandtl number	$Pr(T) = 9.7898\exp(-0.0185T)$	—	0.9756

From: [13].

Various aqueous solutions are widely applied nowadays in industrial and food technologies. The CDP values of some of these solutions are presented in Tables 9.6 to 9.10.

Table 9.6
Dielectric Properties of Saltwater at 915 MHz

Salinity%	$T°C$	20	30	40	50	60	70	80
0.25	ε'	80	77.45	74.24	71.21	68.26	65.48	62.83
	ε''	13.23	13.97	15.24	16.76	18.52	20.42	22.44
0.5	ε'	93.77	88.10	76.83	68.47	64.06	59.151	60.44
	ε''	24.497	26.31	26.77	27.96	30.60	32.64	38.82
0.75	ε'	78.83	76.11	72.36	67.53	62.99	58.68	55.08
	ε''	28.93	32.37	36.39	40.28	44.06	47.64	51.39
1.0	ε'	79.01	76.42	72.66	68.19	63.47	58.53	54.25
	ε''	35.55	40.65	46.47	51.86	56.87	61.22	65.09

From: [14].

Table 9.7
Dielectric Properties of Emulsions and Suspensions at 25°C and $f = 2.45$ GHz

Substance	Volume Fraction	Expression	R^2
Limestone in water	$0 \leq V\% \leq 50$	$\varepsilon'(T) = 0.0031V^2 - 0.9995V + 77.985$	0.9992
		$\varepsilon''(T) = 0.0009V^2 - 0.1642V + 10.182$	0.9851
Oil in water	$0 \leq V\% \leq 80$	$\varepsilon'(T) = -10^{-4}V^3 + 0.0184V^2 - 1.6501V + 75.912$	0.9983
		$\varepsilon''(T) = 2 \cdot 10^{-5}V^3 - 0.0014V^2 - 0.1124V + 11.03$	0.9968
Water in oil	$0 \leq V\% \leq 60$	$\varepsilon'(T) = 8 \cdot 10^{-5}V^3 + 0.0009V^2 + 0.0088V + 2.707$	0.9991
		$\varepsilon''(T) = 4 \cdot 10^{-5}V^3 - 0.0021V^2 + 0.0339V + 0.169$	0.9933

From: [15].

Table 9.8
Dielectric Properties of Synthetic Ballast Water at
$20 \leq T°C \leq 80$

Frequency (MHz)	Expression	R^2
915	$\varepsilon'(T) = 0.0059T^2 - 0.8839T + 99.654$	0.9313
	$\varepsilon''(T) = 54 + 1.935T$	0.9990
2,450	$\varepsilon'(T) = 0.0059T^2 - 0.8342T + 90$	0.9126
	$\varepsilon''(T) = 0.0042T^2 + 0.289T + 30.249$	0.9997

From: [16].

Table 9.9
Dielectric Properties of Naphthenic Acid-Water
Mixture at Concentration 100 ppm

Temperature (°C)	Frequency (MHz)	ε'	ε''
24	915	75.33	6.7
	2,450	75.0	10
35	915	70.26	9.49
	2,450	72.6	7.64

From: [17].

Table 9.10
Dielectric Properties of Water-Silica Gel Solvers at
2,375 MHz $20 \leq T°C \leq 90$

NaCl%	W_w/W_s	Approximate Expression	R^2
0	1.4	$\varepsilon'(T) = 0.0013T^2 - 0.286T + 39.348$	0.9975
		$\varepsilon''(T) = 0.0006T^2 - 0.1574T + 17.231$	0.9981
0	1.5	$\varepsilon'(T) = 48.919 - 0.1976T$	0.9999
		$\varepsilon''(T) = 17.51 - 0.0948T$	0.9981
1.5	1.4	$\varepsilon'(T) = 0.0007T^2 - 0.2065T + 35.678$	0.9985
		$\varepsilon''(T) = 24.609 - 3.5403\ln(T)$	0.9981
1.5	1.5	$\varepsilon'(T) = 43.756 - 0.0899T$	0.9913
		$\varepsilon''(T) = 26.587 - 3.2155\ln(T)$	0.9953

From: [18].

The dielectric properties of supersaturated α-D-glucose at 2.45 GHz in the temperature range $25 \leq T°C \leq 85$ and moisture content $45 \leq W\% \leq 85$ taken from [19]:

$$\varepsilon'(T,W) = u_1 + u_2 T + \frac{u_3}{W} + u_4 T^2 + \frac{u_5}{W^2} + \frac{u_6 T}{W}$$
$$+ u_7 T^3 + \frac{u_8}{W^3} + \frac{u_9 T}{W^2} + \frac{u_{10} T^2}{W}, \quad R^2 = 0.9909 \tag{9.7}$$

$$\varepsilon''(T,W) = u_1 + u_2 T + u_3 W + u_4 T^2 + u_5 W^2 + u_6 TW + u_7 T^3 + u_8 W^3$$
$$+ u_9 TW^2 + u_{10} T^2 W, \quad R^2 = 0.9916 \tag{9.8}$$

where coefficients $u_1 \ldots u_{10}$ are given in Table 9.11.

Table 9.11
Empirical Coefficients of
Equations (9.7) and (9.8)

u_i	ε'	ε''
u_1	517.252	312.419
u_2	3.3201	−0.8456
u_3	−91390.436	−16.693
u_4	−0.006624	0.0154
u_5	$5.20865 \cdot 10^6$	0.34036
u_6	−191.179	−0.01419
u_7	$6.9736 \cdot 10^{-5}$	$-2.759 \cdot 10^{-5}$
u_8	$-9.467 \cdot 10^7$	-0.0023588
u_9	4883.68	0.0003473
u_{10}	−0.4078	−0.0001746

The analogous data of α-D-glucose at 915 MHz can be found in [20].

The thermodynamic properties of mineral salts contained in soils have been measured in [21, 22]. Table 9.12 lists some of the results.

Table 9.12
Dielectric Properties of Two Chemical Substances with Different Humidity
$(0 \leq W \leq 1)$ at 1.11 GHz

Substance	Temperature	Expression
Na_2CO_3 $W = 0.17$	$15 < T°C \leq 80$	$\varepsilon'(T) = (2.68 \pm 0.01) + (1.5 \pm 0.1)\, 10^{-4}\, T$
	$80 < T°C < 97$	$\varepsilon'(T) = (28 \pm 9) - (0.6 \pm 0.2)\, T + (0.004 \pm 0.001)T^2$
	$101 \leq T°C \leq 140$	$\varepsilon'(T) = (1.2 \pm 0.06) + (0.025 \pm 0.001)\, T$
	$15 < T°C \leq 80$	$\varepsilon''(T) = 0.08 \pm 0.007$
	$80 < T°C < 100$	$\varepsilon''(T) = (10 \pm 3) - (0.23 \pm 0.07)\, T + (0.0014 \pm 0.0004)T^2$
	$100 \leq T°C < 140$	$\varepsilon''(T) = (-5.6 \pm 0.7) + (0.1 \pm 0.01)\, T - (3.89 \pm 0.005)T^2$
NaCl $W = 0.01$	$20 \leq T°C \leq 140$	$\varepsilon'(T) = (3.47 \pm 0.01) + (0.0023 \pm 0.0002)T$
		$\varepsilon''(T) = (0.47 \pm 0.008) + (0.0016 \pm 0.0001)T$
NaCl $W = 0.12$	$20 \leq T°C \leq 140$	$\varepsilon'(T) = (7.6 \pm 0.05) - (0.012 \pm 0.001)\, T + (6.4 \pm 0.1) \cdot 10^{-4}T^2$
		$\varepsilon''(T) = (14.4 \pm 0.2) - (0.011 \pm 0.007)\, T - (0.001 \pm 5 \cdot 10^{-5})\, T^2$

From: [21].

As it has been found in [22], moist Na_2CO_3 crystalline hydrates ($W = 0.17$) demonstrates temperature hysteresis of $\varepsilon'(T)–j\varepsilon''(T)$ at 1.11 GHz. In particular, $2.5 \leq \varepsilon'(T) \leq 5.33$ (heating); $2.7 \leq \varepsilon'(T) \leq 6.75$ (cooling) and $0.05 \leq \varepsilon''(T) \leq 1.96$ (heating); $0.06 \leq \varepsilon''(T) \leq 3.11$ (cooling), where $20 \leq T°C \leq 190$.

The dielectric properties of some pure liquids as a function of temperature taken from [23] are given in Tables 9.13 to 9.15.

Table 9.13
Dielectric Properties of Some Chemical Substances in
Microwave Range

Chemical Substance	f (GHz)	T°C	Dielectric Permittivity	Loss Factor
$C_2H_6O_2$	1.43	20	20.38	15.53
		35	27.1	13.77
		50	30.13	9.94
	2.134	10	12.65	12.48
		20	15.85	11.25
		30	20.6	10.9
C_3H_8O	2.97	20	4.35	2.7
		40	5.36	3.5
		60	6.77	5.5
C_4H_4S	2.8	20	2.764	0.013
		40	2.7	0.006
		60	2.63	0.007
C_4H_5N	2.85	20	8.046	0.87
		40	7.67	0.64
		60	7.362	0.47
$C_4H_{10}O_2$	1.43	20	10.06	8.92
		35	13.68	11.16
		50	18.5	10.15
$C_{15}H_{14}O_2$	2	30	3.62	0.91
		50	3.7	0.8
		70	2.92	0.62
CH_4O	1.744	20	25.8	11.6
	3.03	20	18.7	14
C_4H_9N	0.9	20	8.26	0.57
		40	7.28	0.26
		60	6.627	0.16

From: [23].

Table 9.14
Dielectric Properties of Some Chemical Substances at
2 GHz and $T = 28$°C

Substance	$C_4H_{10}O_3$	$C_4H_{10}O$	$C_3H_8O_3$	$C_3H_8O_2$	$C_{15}H_{14}O_2$
Dielectric permittivity	11.9	3.49	9.17	12	3.62
Loss factor	3.2	1.98	6.4	8.8	0.91

From: [23].

Table 9.15
Dielectric Properties of Four Chemical Substances
at 3 GHz

$T°C$	$C_6H_{11}Br$		$C_{10}H_7Cl$	
	Dielectric Permittivity	Loss Factor	Dielectric Permittivity	Loss Factor
25	7.42	1.7	4.16	0.86
40	7.24	1.42	4.22	0.75
55	7.02	1.16	4.29	0.64
75	6.68	0.91	4.35	0.52
$T°C$	C_6H_5Br		$C_6H_5NO_2$	
20	5.09	0.85	21.35	12.90
40	5.02	0.61	21.40	11.60
50	4.92	0.52	23.90	11.15

From: [23].

Table 9.16 lists data on the CDP of two materials used in the polymer industry.

Table 9.16
Dielectric Properties of Polycarbonate and Polyvinyl Chloride at 2.45 GHz

Substance	Temperature	Expression
Polycarbonate	$25 \leq T°C \leq 300$	$\varepsilon'(T) = 7.27 \cdot 10^{-4}T + 4.082$
	$25 \leq T°C < 150$	$\varepsilon''(T) = 0.0055T^2 - 0.4282T + 8.0603$, $R^2 = 0.9887$
	$150 < T°C \leq 300$	$68 \leq \varepsilon''(T) \leq 76$
Polyvinyl chloride	$25 \leq T°C \leq 225$	$\varepsilon'(T) = 5 \cdot 10^{-5}T^2 - 0.0194T + 4.4552$, $R^2 = 0.9998$
	$25 \leq T°C < 100$	$0 \leq \varepsilon''(T) \leq 0.1$
	$100 < T°C < 150$	$0.1 \leq \varepsilon''(T) \leq 1.7$
	$150 < T°C \leq 225$	$1.7 \leq \varepsilon''(T) \leq 1.9$

From: [24].

Different chemical substances are widely employed in pharmacology for manufacturing medical drugs. Microwaves radiation accelerates synthesis of some substances. Dielectric and thermal properties of some liquids and powders used for these purposes have been measured and published in [4].

Table 9.17
Dielectric Properties of Liquids and Powders at 2.45 GHz
Used in Pharmacology

Matter	$T°C$	20	30	40	50	60	70
Ethanol	ε'	7.49	8.01	11.16	12.76	14.48	14.72
	ε''	6.73	7.17	8.8	8.89	8.6	8.1
Acetone	ε'	23.36	21.69	21.22	20.72	19.54	19.1
	ε''	1.29	0.85	0.81	0.83	0.57	0.55
Paracetamol powder (Pa)	ε'	1.92	6.32	8.94	10.21	10.98	12.6
	ε''	0.07	0.36	0.58	0.73	0.84	1.07
Aspirin powder (As)	ε'	1.56	4.38	6.71	7.76	8.05	11.85
	ε''	0.05	0.31	0.49	0.53	0.77	1.23
Lactose powder (La)	ε'	1.98	6.99	9.5	13.36	15.77	—
	ε''	0.07	1.11	1.7	1.83	2.0	—
Adipic acid powder (Ad)	ε'	2.05	3.32	5.11	5.88	7.31	8.8
	ε''	0.09	0.46	0.65	0.81	1.18	1.43

From: [4].

Table 9.18
Thermal Properties of Pharmaceutical Powders and Liquids at Room Temperature

Matter	Ethanol	Acetone	Paracetamol	Aspirin	Lactose	Adipic Acid
C_t (kJ/kg·K)	1.98	1.65	2.41	2.2	2.31	2.16
λ_t (W/m·K)	0.17	0.16	0.12	0.1	0.19	0.15
a_t (mm²/s)	0.05	0.1	0.05	0.08	0.08	0.07

From: [4].

Table 9.19
Dielectric Properties of Pharmaceutical Lactose-Solvent Mixtures at 2.45 GHz

Solvent	W%	Temperature (°C) 20 ε'	20 ε''	30 ε'	30 ε''	40 ε'	40 ε''	50 ε'	50 ε''	60 ε'	60 ε''
Water	30	13.94	1.47	16.26	1.94	21.49	2.45	24.8	2.77	28.8	2.93
	50	19.68	2.29	25.73	3.17	32	3.36	34.11	3.71	42.17	4.2
Ethanol	30	2.79	1.06	6.0	1.59	7.61	2.51	8.97	3.11	10.01	3.8
	50	3.99	3.24	5.24	3.52	5.99	4.24	6.49	5.12	6.73	5.32
Acetone	30	4.43	0.29	9.2	0.99	11.26	1.06	14.36	1.69	16.13	1.86
	50	9.1	0.52	13.15	0.71	15.22	0.92	15.92	1.2	16.51	1.41

From: [4].

Table 9.20

Dielectric Properties of Pharmaceutical Powders-Solvent Mixtures at 2.45 GHz

Powder	Solvent	W%	T°C	20	30	40	50	60	70
Paracetamol	Water	30	ε'	21.64	—	25.29	31.47	31.91	37.51
			ε''	1.84	2.83	3.04	3.83	3.95	4.66
		50	ε'	26.58	27.77	32.7	36.34	44.73	48.31
			ε''	2.83	3.36	4.09	4.52	5.61	5.94
	Ethanol	30	ε'	2.72	3	8.71	9.65	10.17	11.21
			ε'	1.65	2.11	2.75	3.11	3.69	4.12
		50	ε'	4.01	4.5	8.43	8.95	9.87	10.12
			ε''	3.02	3.5	4.12	4.5	4.91	5.46
	Acetone	30	ε'	10.42	10.83	11.11	11.73	12.57	13.9
			ε'	0.51	0.54	0.72	0.89	0.98	1.47
		50	ε''	12.27	11.86	10.98	10.65	10.35	10.01
			ε'	0.79	0.72	0.7	0.68	0.65	0.62
Aspirin	Water	30	ε'	17.69	25.19	30.8	36.56	39.4	43.74
			ε''	2.07	2.94	3.9	4.58	4.97	5.88
		50	ε'	30.72	35.03	43.94	44.7	46.43	48.97
			ε''	3.16	4.51	5.74	5.81	5.91	6.46
	Ethanol	30	ε'	3.02	4.9	7.86	8.72	9.42	12.32
			ε''	1.82	2.02	2.42	2.9	3.36	3.72
		50	ε'	4.12	5.25	8.41	9.57	9.99	12.92
			ε''	3.21	3.62	3.97	4.34	4.93	5.2
	Acetone	30	ε'	5.85	5.57	7.53	7.72	8.01	8.71
			ε''	0.33	0.39	0.58	0.66	0.89	1.72
		50	ε'	11.29	10.7	10.29	10.0	9.62	9.65
			ε''	0.69	0.66	0.64	0.63	0.59	0.5
Adipic acid	Water	30	ε'	21.28	25.2	31.29	30.36	36.16	38.79
			ε''	2.36	3.01	3.2	3.72	4.04	4.46
		50	ε'	30.66	35.48	38.5	40.93	43.37	47.47
			ε''	3.4	3.67	4.05	4.49	5.49	5.7
	Ethanol	30	ε'	3.11	3.78	5.12	6.14	7.91	8.9
			ε''	1.93	2.17	2.49	2.71	3.0	3.27
		50	ε'	4.13	4.5	5.13	6.38	8.19	10.91
			ε''	3.06	3.41	3.71	3.91	4.12	4.32
	Acetone	30	ε'	7.3	5.72	5.17	4.33	6.83	6.85
			ε''	0.31	0.53	0.58	0.64	0.69	0.4
		50	ε'	12.24	12.0	11.83	11.52	11.12	12.23
			ε''	0.6	0.57	0.54	0.63	0.62	0.59

From: [4].

Some chemical substances are successfully applied in medicine for phantom materials design, which are used in experimental studies of processes of microwaves propagation and attenuation in biological tissues. Some of these materials are shown in Table 9.21. Table 9.22 lists data on the CDP of solutions also proposed for phantom materials design. Additional information about the CDP of various chemical substances can be found in [27, 28] (see Tables 9.23 and 9.24).

Table 9.21
Dielectric Properties of Phantom Materials Used in Microwave Medicine at $T = 22°C$

Tissue	Composition	f (MHz)	ε'	ε''
Bone and fat phantom	82.5% laminac 4110, 14.5% Al powder; 0.24% acetylene black, 3.75 g/kg peroxide	915	5.6	1.3
		2,450	4.5	0.84
Brain phantom	62.61% H_2O, 0.58% NaCl, 29.8% polyethylene powder, 7.01% super stuff	915	34.4	15.2
		2,450	33.6	9.1
	59% H_2O, 1% NaCl, 40% gelatin	2,450	42	19
Muscle phantom	75.4% H_2O, 0.907% NaCl, 15.2% polyethylene powder, 8.45% super stuff	915	50.6	26.6
		2,450	49.6	16.5
	69% H_2O, 30% gelatin, 1% NaCl	2,450	50	16
Skin phantom	60% H_2O, 1% NaCl, 39% cellulose paper	2,450	43	15

From: [25].

Table 9.22
Dielectric Properties of the Sodium Chloride Solutions at 430 MHz
in Temperature Range $20 \leq T°C \leq 60$

0.2N	$\varepsilon'(T) = 81.4 - 0.19T$	$\varepsilon''(T) = 46.46 + 1.559T$
0.4N	$\varepsilon'(T) = 80.286 - 0.1028T$	$\varepsilon''(T) = 92.696 + 2.6505T$
0.6N	$\varepsilon'(T) = 79.995 - 0.01475T$	$\varepsilon''(T) = 216.95 + 2.753T$
0.8N	$\varepsilon'(T) = 76.56 + 0.172T$	$\varepsilon''(T) = 167.78 + 5.211T$
1.0N	$\varepsilon'(T) = 75.645 + 0.335T$	$\varepsilon''(T) = 209.76 + 7.011T$

From: [26].
N=normal concentration of NaCl sodium chloride solution.

Table 9.23

Dielectric Properties of N,N-Dimethylformamide in Temperature Range $20 \le T°C \le 80$

Frequency	Expression	R^2
2 GHz	$\varepsilon'(T) = 6 \cdot 10^{-5}T^3 - 0.0102T^2 + 0.347T + 34.239$	0.9983
	$\varepsilon''(T) = 5.5888 - 0.0468T$	0.9873
2.45 GHz	$\varepsilon'(T) = 7 \cdot 10^{-5}T^3 - 0.0106T^2 + 0.3289T + 35.98$	0.9983
	$\varepsilon''(T) = 7.169 - 0.0595T$	0.9776
3 GHz	$\varepsilon'(T) = 2 \cdot 10^{-5}T^3 - 0.0038T^2 + 0.0789T + 38.634$	0.9991
	$\varepsilon''(T) = 8.788 - 0.0656T$	0.9931

From: [27].

Table 9.24

Dielectric Properties of Organic Liquids at 2,450 MHz and Temperature $T = 25°C$

Liquid	ε'	ε''	Liquid	ε'	ε''
H_2O	78.6	10.11	$(ClCH_2)_2$	10.1	0.91
DMSO	43.0	10.86	$1,2-Cl_2C_6H_4$	9.0	2.3
$MeNO_2$	36.5	2.24	CH_2Cl_2	8.7	0.21
MeCN	35.8	1.7	EtOH	6.8	6.89
NMI	30.7	11.36	MeI	6.7	0.25
NMP	29.4	7.72	$MeCO_2Et$	5.9	0.25
$PhNO_2$	23.3	14.39	PhCl	5.4	0.53
MeOH	21.3	13.14	$CHCl_3$	4.6	0.22
Me_2CO	20.3	0.86	Et_2O	4.1	0.075
PhCN	20.1	8.73	BuOH	3.7	1.98
MeCOEt	17.8	1.11	MePh	2.4	0.012
PhCHO	15.5	5.02	PhH	2.2	0.00084
$[OCN(CH_2)_3]_2$	11.9	4.8	hexane	1.9	0.00078

From: [28].

Sometimes one can find information in the literature about the physical properties of very specific materials such as a thermally regenerable chemisorbent for carbon dioxide that is used for air purification in spacecrafts (see Table 9.25).

Table 9.25
Dielectric Properties of Ag_2O-Ag_2CO_3 System at 6 GHz

Material	Temperature	Dielectric Permittivity	Loss Factor
Silver oxide	22.5°C	1.69	0.18
	50.5°C	1.64	0.12
	188.8°C	1.67	0.125
Silver carbonate	22.7°C	2.45	0.33
	50.6°C	2.35	0.14
	99.2°C	2.4	0.21
	150.8°C	2.39	0.12
	188.6°C	2.05	0.13

From: [29].

Today, new artificial composites and different nanomaterials are being used for numerous practical applications in science and engineering and are being intensively studied. The CDP of such nanocomposites have been described in many publications [29–34].

Several books and papers devoted to investigating the thermophysical properties of chemical substances contain data about only one parameter of this or that matter, but in many publications one can find information about all three main parameters (thermal conductivity, heat capacity, and density) as shown in Tables 9.26 to 9.50. Moreover, the data found on kinematic (ν_t) or dynamic (μ_t) viscosity for liquids can be useful in many cases.

Table 9.26
Thermal Properties of Heavy Water at Pressure 1 bar $10 \leq T°C \leq 90$

Parameter	Expression	R^2	Units
Density	$\rho_t(T) = 3 \cdot 10^{-5}T^3 - 0.0083T^2 + 0.1683T + 1105.1$	0.9999	kg/m³
Heat capacity	$C_t(T) = 3 \cdot 10^{-4}T^3 - 0.0624T^2 + 2.4403T + 4213.6$	0.9983	J/kg·K
Thermal conductivity	$\lambda_t(T) = -4 \cdot 10^{-6}T^2 + 0.0011T + 0.5594$	0.9998	W/m·K
Kinematical viscosity ×10^{-6}	$\nu_t(T) = -4 \cdot 10^{-6}T^3 + 0.0008T^2 - 0.0588T + 2.0563$	0.9990	m²/s

From: [35].

Table 9.27
Thermal Parameters of Some Organic Liquids

Liquid	Parameter	$T = 0°C$	$T = 20°C$	$T = 50°C$	Units
Methanol CH_4O	$\rho_t(T)$	810	792	765	kg/m^3
	$C_t(T)$	2.386	2.495	2.68	kJ/kg·K
	$\lambda_t(T)$	0.208	0.202	0.193	W/m·K
	$\mu_t(T) \times 10^{-5}$	81.7	58.4	3.96	Pa·s
Ethanol C_2H_6O	$\rho_t(T)$	806	789	763	kg/m^3
	$C_t(T)$	2.232	2.395	2.801	kJ/kg·K
	$\lambda_t(T)$	0.177	0.173	0.165	W/m·K
	$\mu_t(T) \times 10^{-5}$	1768.6	120.1	70.1	Pa·s
Glycerol $C_3H_8O_3$	$20 \leq T°C \leq 200$			R^2	Units
	$\rho_t(T) = -0.0015T^2 - 1.1212T + 1289.2$			0.9939	kg/m^3
	$C_t(T) = 6 \cdot 10^{-8}T^3 - 2 \cdot 10^{-5}T^2 + 0.0069T + 2.225$			0.9994	kJ/kg·K
	$\lambda_t(T) = -6 \cdot 10^{-9}T^3 + 10^{-6}T^2 + 0.00009T + 0.2839$			0.9999	W/m·K
	$\mu_t(T) \times 10^{-5} = 211836\exp(-0.0459)$			0.9908	Pa·s

From: [35].

Table 9.28
Thermal Conductivity of Some Pure Liquids

Liquid	Thermal Conductivity (W/m·K)	Temperature
Acetone	0.171	$T = 273°K$
Acetone-nitrile	$0.2307 \leq \lambda_t(T) \leq 0.2168$	$253 \leq T°K \leq 293$
Chloroform	$0.109 \leq \lambda_t(T) \leq 0.121$	$270 \leq T°K \leq 340$
N-decyl alcohol	$0.147 \leq \lambda_t(T) \leq 0.162$	$298 \leq T°K \leq 398$
N-licosane	$0.118 \leq \lambda_t(T) \leq 0.147$	$330 \leq T°K \leq 460$
Ethanol	$0.143 \leq \lambda_t(T) \leq 0.196$	$200 \leq T°K \leq 400$
Heptane	$\lambda_t(T) = 2 \cdot 10^{-7}T^2 - 0.0004T + 0.2233,$ $R^2 = 0.9996$	$200 \leq T°K \leq 500$
Ethylbenzene	$0.1044 \leq \lambda_t(T) \leq 0.1369$	$237 \leq T°K \leq 400$
Toluene	$0.1125 \leq \lambda_t(T) \leq 0.1485$	$240 \leq T°K \leq 360$

From: [36].

Table 9.29
Thermal Conductivity (W/m·K) of Some Inorganic Liquids

Substance	T°C	Expression	Concentration
KOH	1.5	$\lambda_t(D) = 10^{-6}D^3 - 10^{-4}D^2 + 0.0021D + 0.5638,$ $R^2 = 0.9961$	$0 \leq D\% \leq 40$
	20	$\lambda_t(D) = 7 \cdot 10^{-7}D^3 - 8 \cdot 10^{-5}D^2 + 0.0015D + 0.5968,$ $R^2 = 0.9997$	
	50	$\lambda_t(D) = 5 \cdot 10^{-7}D^3 - 7 \cdot 10^{-5}D^2 + 0.0011D + 0.6401,$ $R^2 = 0.9993$	
	80	$\lambda_t(D) = 7 \cdot 10^{-7}D^3 - 9 \cdot 10^{-5}D^2 + 0.0014D + 0.6691,$ $R^2 = 0.9999$	
NaOH	1.5	$\lambda_t(D) = 3 \cdot 10^{-6}D^3 - 2 \cdot 10^{-4}D^2 + 0.0053D + 0.5632,$ $R^2 = 0.9993$	$0 \leq D\% \leq 30$
	20	$\lambda_t(D) = -7 \cdot 10^{-5}D^2 + 0.0034D + 0.6,$ $R^2 = 0.9946$	
	50	$\lambda_t(D) = 10^{-6}D^3 - 10^{-4}D^2 + 0.0048D + 0.6389,$ $R^2 = 0.9899$	
	80	$\lambda_t(D) = 10^{-6}D^3 - 10^{-4}D^2 + 0.0047D + 0.6691,$ $R^2 = 0.9992$	
NaCl	1.5	$\lambda_t(D) = -0.0007D + 0.5629,\ R^2 = 0.9984$	$0 \leq D\% \leq 25$
	20	$\lambda_t(D) = -10^{-5}D^2 - 0.0007D + 0.599,\ R^2 = 0.9998$	
	25	$\lambda_t(D) = -10^{-5}D^2 - 0.0007D + 0.6084,\ R^2 = 0.9981$	
	30	$\lambda_t(D) = -0.0014D + 0.618,\ R^2 = 0.9999$	

From: [37].

Table 9.30
Thermal Conductivity (W/m·K) of Some Organic Liquids

Substance	T°C	Expression	Concentration
C_2H_6O	20	$\lambda_t(D) = 3 \cdot 10^{-5}D^2 - 0.0071D + 0.5981,\ R^2 = 0.9999$	$0 \leq D\% \leq 80$
	60	$\lambda_t(D) = 3 \cdot 10^{-5}D^2 - 0.0081D + 0.6521,\ R^2 = 0.9999$	
$C_3H_8O_3$	10	$\lambda_t(D) = 8 \cdot 10^{-6}D^2 - 0.0038D + 0.5817,\ R^2 = 0.9997$	$0 \leq D\% \leq 90$
	20	$\lambda_t(D) = 7 \cdot 10^{-6}D^2 - 0.0039D + 0.5991,\ R^2 = 0.9999$	
	50	$\lambda_t(D) = -0.0037D + 0.6409,\ R^2 = 0.9999$	
	80	$\lambda_t(D) = 6 \cdot 10^{-6}D^2 - 0.0045D + 0.6709,\ R^2 = 0.9999$	
C_3H_6O	20	$\lambda_t(D) = 3 \cdot 10^{-5}D^2 - 0.0074D + 0.5991,\ R^2 = 0.9999$	$0 \leq D\% \leq 80$
	40	$\lambda_t(D) = 3 \cdot 10^{-5}D^2 - 0.008D + 0.6277,\ R^2 = 0.9999$	
$C_2H_{12}O_6$ and $C_{12}H_{22}O_{11}$	1.5	$\lambda_t(D) = -4 \cdot 10^{-7}D^2 - 0.0029D + 0.5668,\ R^2 = 0.9994$	$0 \leq D\% \leq 60$
	20	$\lambda_t(D) = -5 \cdot 10^{-7}D^2 - 0.0032D + 0.5999,\ R^2 = 0.9998$	
	50	$\lambda_t(D) = -0.0035D + 0.6415,\ R^2 = 0.9995$	
	80	$\lambda_t(D) = -6 \cdot 10^{-6}D^2 - 0.0034D + 0.6726,\ R^2 = 0.9998$	

From: [37].

Table 9.31
Density of Aqueous Solutions of Some Chemical Substances

Substance	Temperature	Concentration	Density (g/cm^3)
KOH	$0 \leq T°C \leq 100$	$2 \leq D\% \leq 50$	$0.9765 \leq \rho_t \leq 1.5257$
NH$_3$		$2 \leq D\% \leq 45$	$0.808 \leq \rho_t \leq 0.9919$
NaOH		$2 \leq D\% \leq 50$	$0.9796 \leq \rho_t \leq 1.54$
HCl		$2 \leq D\% \leq 35$	$0.968 \leq \rho_t \leq 1.1875$
HNO$_3$		$2 \leq D\% \leq 95$	$0.9685 \leq \rho_t \leq 1.5495$
H$_2$SO$_4$		$2 \leq D\% \leq 95$	$0.9705 \leq \rho_t \leq 1.8511$
C$_3$H$_5$(OH)$_3$	$0 \leq T°C \leq 60$	$5 \leq D\% \leq 75$	$0.995 \leq \rho_t \leq 1.1899$
C$_{12}$H$_{22}$O$_{11}$	$0 \leq T°C \leq 100$	$2 \leq D\% \leq 45$	$0.965 \leq \rho_t \leq 1.1834$
HCOOH	$0 \leq T°C \leq 40$	$2 \leq D\% \leq 40$	$0.9966 \leq \rho_t \leq 1.1094$
CH$_3$COOH	$0 \leq T°C \leq 60$	$2 \leq D\% \leq 90$	$0.985 \leq \rho_t \leq 1.0863$
CH$_3$COCH$_3$	$0 \leq T°C \leq 60$	$10 \leq D\% \leq 90$	$0.799 \leq \rho_t \leq 0.99$

From: [37].

Table 9.32
Heat Capacity of Some Liquids

Liquid	Temperature	Heat capacity (kJ/kg·K)	R^2
C$_3$H$_6$O	$0 \leq T°C \leq 50$	$C_t(T) = 2 \cdot 10^{-5}T^2 + 0.0017T + 2.119$	0.9999
C$_6$H$_7$N	$0 \leq T°C \leq 150$	$C_t(T) = 5 \cdot 10^{-5}T^2 - 0.0027T + 2.1045$	0.9947
C$_2$H$_6$O	$0 \leq T°C \leq 160$	$C_t(T) = 7 \cdot 10^{-5}T^2 + 0.0039T + 2.3483$	0.9937
C$_2$H$_6$O$_2$	$20 \leq T°C \leq 100$	$C_t(T) = 0.0045T + 2.2932$	0.9999
C$_2$H$_5$J	$0 \leq T°C \leq 60$	$C_t(T) = 0.0007T + 0.6769$	0.9992
C$_6$H$_6$	$10 \leq T°C \leq 65$	$C_t(T) = 2 \cdot 10^{-5}T^3 - 0.0021T^2 + 0.0801T + 0.8229$	0.9941
C$_6$H$_5$Br	$20 \leq T°C \leq 80$	$C_t(T) = 10^{-5}T^2 - 0.0003T + 0.9683$	0.9999
C$_4$H$_{10}$O	$10 \leq T°C \leq 85$	$C_t(T) = 0.0189T + 1.9285$	0.9989
C$_6$H$_5$Cl	$20 \leq T°C \leq 80$	$C_t(T) = 5 \cdot 10^{-6}T^2 + 0.0003T + 0.9585$	0.9999
C$_{10}$H$_3$	$90 \leq T°C \leq 190$	$C_t(T) = 0.0032T + 1.4892$	0.9999
C$_{10}$H$_{16}$	$0 \leq T°C \leq 100$	$C_t(T) = 10^{-5}T^2 + 0.0035T + 1.7193$	0.9994
C$_7$H$_8$	$0 \leq T°C \leq 100$	$C_t(T) = -5 \cdot 10^{-7}T^3 + 8 \cdot 10^{-5}T^2 + 0.0008T + 1.633$	0.9999

From: [38].

Table 9.33
Thermal properties of liquid H_2O_2 in temperature range
$0 \leq T°C \leq 160$

Units	Parameter	R^2
kg/m³	$\rho_t(T) = 1474.8 - 1.2033T$	0.9995
W/m·K	$\lambda_t(T) = -4 \cdot 10^{-6}T^2 + 0.0014T + 0.453$	0.9999
J/kg·K	$C_t(T) = 2.07T + 2599$	0.9999
mPa·s	$\mu_t(T) = 1.5517\exp(-0.0108T)$	0.9883

From: [39].

Table 9.34
Density of Some Liquids

Liquid	Temperature	Density (kg/m³)	R^2
$C_4H_8O_2$	$0 \leq T°C \leq 240$	$\rho_t(T) = -3 \cdot 10^{-5}T^3 + 0.0059T^2 - 1.6156T + 929.12$	0.9987
C_2H_6O	$0 \leq T°C \leq 240$	$\rho_t(T) = -4 \cdot 10^{-5}T^3 + 0.0082T^2 - 1.3907T + 812.85$	0.9959
C_2H_7N	$0 \leq T°C \leq 80$	$\rho_t(T) = -1.2567T + 708.04$	0.9998
C_2H_5Cl	$0 \leq T°C \leq 80$	$\rho_t(T) = -0.0028T^2 - 1.3317T + 919.77$	0.9998
$C_2H_6O_2$	$20 \leq T°C \leq 100$	$\rho_t(T) = -0.715T + 1127.5$	0.9999
C_6H_6	$0 \leq T°C \leq 280$	$\rho_t(T) = -2 \cdot 10^{-5}T^3 + 0.0051T^2 - 1.4545T + 905.7$	0.9983
C_6H_5Br	$0 \leq T°C \leq 270$	$\rho_t(T) = -0.0011T^2 - 1.1833T + 1518.3$	0.9997
Cl_2	$0 \leq T°C \leq 130$	$\rho_t(T) = -0.0146T^2 - 2.1222T + 1459.1$	0.9981
C_6H_5Cl	$0 \leq T°C \leq 270$	$\rho_t(T) = -0.0013T^2 - 0.9016T + 1124.2$	0.9995
$C_4H_{10}O$	$0 \leq T°C \leq 140$	$\rho_t(T) = -0.0031T^2 - 0.9452T + 734.13$	0.9993
CH_5N	$10 \leq T°C \leq 80$	$\rho_t(T) = -1.4762T + 690.68$	0.9991
$C_2H_4O_2$	$0 \leq T°C \leq 300$	$\rho_t(T) = -10^{-5}T^3 + 0.0028T^2 - 1.3127T + 1074.4$	0.9994
$C_3H_6O_2$	$0 \leq T°C \leq 220$	$\rho_t(T) = -3 \cdot 10^{-5}T^3 + 0.0051T^2 - 1.6003T + 962.68$	0.9993
CH_4O	$0 \leq T°C \leq 220$	$\rho_t(T) = -2 \cdot 10^{-5}T^3 + 0.0038T^2 - 1.1089T + 812.31$	0.9994
$C_5H_{10}O_2$	$0 \leq T°C \leq 270$	$\rho_t(T) = -2 \cdot 10^{-5}T^3 + 0.0041T^2 - 1.4018T + 913.75$	0.9993
C_7H_8	$20 \leq T°C \leq 200$	$672 \leq \rho_t \leq 866$	

From: [38].

Table 9.35
Dynamic Viscosity of Some Liquids

Liquid	Temperature	Dynamic Viscosity $\cdot 10^4$ (Pa$\cdot$s)	R^2
C_3H_6O	$0 \leq T°C \leq 50$	$\mu_t(T) = 2 \cdot 10^{-4}T^2 - 0.0407T + 3.9874$	0.9935
C_2H_6O	$0 \leq T°C \leq 150$	$\mu_t(T) = -7 \cdot 10^{-6}T^3 + 0.0024T^2 - 0.3185T + 17.638$	0.9994
$C_2H_6O_2$	$20 \leq T°C \leq 140$	$\mu_t(T) = 2577T^{-1.561}$	0.9901
C_6H_6	$0 \leq T°C \leq 190$	$\mu_t(T) = -2 \cdot 10^{-6}T^3 + 0.0008T^2 - 0.1247T + 8.8417$	0.9975
$C_3H_6O_2$	$20 \leq T°C \leq 140$	$\mu_t(T) = 4.4557\exp(-0.009T)$	0.9991
C_4HO	$0 \leq T°C \leq 70$	$\mu_t(T) = -9 \cdot 10^{-6}T^3 + 0.0017T^2 - 0.1446T + 8.1645$	0.9998
$C_{10}H_8$	$80 \leq T°C \leq 150$	$\mu_t(T) = 20.487\exp(-0.01T)$	0.9986
$C_5H_{10}O_2$	$0 \leq T°C \leq 100$	$\mu_t(T) = -4 \cdot 10^{-6}T^3 + 0.0009T^2 - 0.1105T + 7.6964$	0.9998
C_5H_5N	$0 \leq T°C \leq 90$	$\mu_t(T) = -7 \cdot 10^{-6}T^3 + 0.002T^2 - 0.2151T + 13.258$	0.9998
CCl_4	$0 \leq T°C \leq 180$	$\mu_t(T) = -3 \cdot 10^{-6}T^3 + 0.0012T^2 - 0.1817T + 13.063$	0.9989
C_7H_8	$0 \leq T°C \leq 140$	$\mu_t(T) = -2 \cdot 10^{-6}T^3 + 0.0007T^2 - 0.0986T + 7.6226$	0.9996

From: [38].

Table 9.36
Thermal Conductivity of Some Liquids

Liquid	Temperature	Thermal conductivity (W/m$\cdot$K)
C_3H_6O	$0 \leq T°C \leq 100$	$0.166 \leq \lambda_t \leq 0.184$
C_6H_7N	$20 \leq T°C \leq 150$	$0.158 \leq \lambda_t \leq 0.172$
C_2H_6O	$0 \leq T°C \leq 75$	$\lambda_t(T) = -0.0001T + 0.1848$, $R^2 = 0.9938$
$C_2H_6O_2$	$0 \leq T°C \leq 100$	$\lambda_t(T) = 0.0001T + 0.2551$, $R^2 = 0.9998$
C_6H_6	$20 \leq T°C \leq 80$	$0.151 \leq \lambda_t \leq 0.154$
C_6H_5Br	$0 \leq T°C \leq 100$	$0.111 \leq \lambda_t \leq 0.13$
$C_3H_8O_3$	$0 \leq T°C \leq 100$	$0.283 \leq \lambda_t \leq 0.291$
CH_4O	$0 \leq T°C \leq 100$	$0.184 \leq \lambda_t \leq 0.214$
C_7H_8	$0 \leq T°C \leq 100$	$0.137 \leq \lambda_t \leq 0.151$

From: [38].

Table 9.37
Thermal Parameters of Some Liquids

Liquid	Parameter	Units
26.5% $C_{12}H_{10}$ + 73.5% $C_{12}H_{10}O$	$C_t(T) = 3 \cdot 10^{-8}T^3 - 2 \cdot 10^{-5}T^2 + 0.0057T + 1.4496$, $R^2 = 0.9995$ $20 \leq T°C \leq 400$	kJ/kg·K
	$\rho_t(T) = -0.8715T + 1082.3$, $R^2 = 0.9996$ $20 \leq T°C \leq 300$	kg/m³
	$\lambda_t(T) = -0.0001T + 0.1416$, $R^2 = 0.9990$ $20 \leq T°C \leq 300$	W/m·K
Aero engine oil	$C_t(T) = 0.0043T + 1.7466$, $R^2 = 0.9995$ $20 \leq T°C \leq 140$	kJ/kg·K
	$\lambda_t(T) = -9 \cdot 10^{-5}T + 0.1467$, $R^2 = 0.9912$ $20 \leq T°C \leq 140$	W/m·K
	$\mu_t(T) \cdot 10^4 = 10386\exp(-0.0397T)$, $R^2 = 0.9592$ $20 \leq T°C \leq 140$	Pa·s
Spindle oil	$C_t(T) = 0.0042T + 1.767$, $R^2 = 0.9999$ $20 \leq T°C \leq 120$	kJ/kg·K
	$\rho_t(T) = -0.6386T + 883.53$, $R^2 = 0.9999$ $20 \leq T°C \leq 120$	kg/m³
	$0.138 \leq \lambda_t \leq 0.144$, $0 \leq T°C \leq 120$	W/m·K
	$\mu_t(T) \cdot 10^4 = 168.11\exp(-0.0211T)$, $R^2 = 0.9758$, $20 \leq T°C \leq 120$	Pa·s

From: [38].

Table 9.38
Thermal Properties of Aqueous Solutions of NaCl

Temperature	Parameter	R^2	Units
$0 \leq T°C \leq 100$	$\rho_t(T) = -0.0045T^2 - 0.0421T + 1018.8$, $D = 2.5\%$	0.9996	kg/m³
	$\rho_t(T) = -0.0022T^2 - 0.2778T + 1077.1$, $D = 10\%$	0.9996	
	$\rho_t(T) = -0.0022T^2 - 0.2778T + 1077.1$, $D = 20\%$	0.9998	
	$\lambda_t(T) = -7 \cdot 10^{-6}T^2 + 0.0019T + 0.5623$, $D = 2.5\%$	0.9999	W/m·K
	$\lambda_t(T) = -7 \cdot 10^{-6}T^2 + 0.0019T + 0.5564$, $D = 10\%$	0.9998	
	$\lambda_t(T) = -7 \cdot 10^{-6}T^2 + 0.0018T + 0.5453$, $D = 20\%$	0.9999	
$0 \leq T°C \leq 30$	$C_t(T) = 1.0429T + 4039.8$, $D = 2.5\%$	0.9994	J/kg·K
	$C_t(T) = T + 3708$, $D = 10\%$	0.9999	
	$C_t(T) = 0.9143T + 3407.9$, $D = 20\%$	0.9991	
$0 \leq T°C \leq 100$	$\mu_t(T) = -3 \cdot 10^{-6}T^3 + 0.0006T^2 - 0.0466T + 1.7789$, $D = 2.5\%$	0.9975	mPa·s
	$\mu_t(T) = -3 \cdot 10^{-6}T^3 + 0.0006T^2 - 0.0499T + 1.9962$, $D = 10\%$	0.9978	
	$\mu_t(T) = -4 \cdot 10^{-6}T^3 + 0.0008T^2 - 0.0682T + 2.6334$, $D = 20\%$	0.9976	

From: [39].

Table 9.39
Thermal Properties of Aqueous Solutions of Na_2CO_3

Temperature	Parameter	R^2	Units
$0 \leq T°C \leq 100$	$\rho_t(T) = -0.0034T^2 - 0.1225T + 1027.4,\ D = 2.5\%$	0.9997	kg/m^3
	$\rho_t(T) = -0.0021T^2 - 0.3224T + 1110.3,\ D = 10\%$	0.9999	
	$\rho_t(T) = -0.6794T + 1229.1,\ D = 20\%$	0.9994	
	$\lambda_t(T) = -8 \cdot 10^{-6}T^2 + 0.0019T + 0.5651,\ D = 2.5\%$	0.9999	W/m·K
	$\lambda_t(T) = -8 \cdot 10^{-6}T^2 + 0.0019T + 0.5721,\ D = 10\%$	0.9999	
	$\lambda_t(T) = -8 \cdot 10^{-6}T^2 + 0.0019T + 0.5808,\ D = 20\%$	0.9999	
$15 \leq T°C \leq 100$	$C_t(T) = 0.0042T^2 + 0.8783T + 4036.9\ ,\ D = 2.5\%$	0.9999	J/kg·K
	$C_t(T) = -6 \cdot 10^{-4}T^3 + 0.1222T^2 - 4.4509T + 3849.2,\ D = 10\%$	0.9999	
	$C_t(T) = -3 \cdot 10^{-4}T^3 + 0.0993T^2 - 5.5176T + 3698.2,\ D = 20\%$	0.9999	
$30 \leq T°C \leq 100$	$\mu_t(T) = 0.0001T^2 - 0.0288T + 1.8576,\ D = 2.5\%$	0.9996	mPa·s
	$\mu_t(T) = 0.0002T^2 - 0.0442T + 2.7119,\ D = 10\%$	0.9994	
	$\mu_t(T) = 0.0005T^2 - 0.105T + 6.0368,\ D = 20\%$	0.9991	

From: [39].

Table 9.40
Thermal Properties of Aqueous Solutions of Saccharose
in Temperature Range $20 \leq T°C \leq 100$

D%	Parameter	R^2	Units
10	$\rho_t(T) = -0.0032T^2 - 0.1433T + 1042.5$	0.9993	kg/m^3
	$C_t(T) = -3 \cdot 10^{-5}T^3 + 0.0216T^2 - 1.8486T + 3978.9$	0.9993	J/kg·K
	$\lambda_t(T) = -6 \cdot 10^{-6}T^2 + 0.0017T + 0.5333$	0.9993	W/m·K
	$\mu_t(T) = -2 \cdot 10^{-6}T^3 + 0.0005T^2 - 0.0467T + 2.0706$	0.9995	mPa·s
30	$\rho_t(T) = 4 \cdot 10^{-5}T^3 - 0.008T^2 - 0.0106T + 1129.8$	0.9987	kg/m^3
	$C_t(T) = -3 \cdot 10^{-5}T^3 + 0.0217T^2 - 1.873T + 3502.4$	0.9995	J/kg·K
	$\lambda_t(T) = -5 \cdot 10^{-6}T^2 + 0.0015T + 0.4735$	0.9999	W/m·K
	$\mu_t(T) = -3 \cdot 10^{-6}T^3 + 0.0009T^2 - 0.1012T + 4.9461$	0.9982	mPa·s
60	$\rho_t(T) = -0.0006T^2 - 0.4761T + 1296.2$	0.9995	kg/m^3
	$C_t(T) = -3 \cdot 10^{-5}T^3 + 0.0216T^2 - 1.8486T + 2785.9$	0.9993	J/kg·K
	$\lambda_t(T) = -4 \cdot 10^{-6}T^2 + 0.0012T + 0.3824$	0.9998	W/m·K
	$\mu_t(T) = 92.678\exp(-0.0354T)$	0.9831	mPa·s

From: [39].

The thermal conductivity of nitric and sulfuric acids [37] is:

$$\lambda_t(D,T) = q_1 + q_2 D + q_3 T + q_4 D^2 + q_5 T^2 + q_6 DT$$
$$+ q_7 D^3 + q_8 DT^2 + q_{10} D^2 T,\ \ R^2 = 0.9996 \tag{9.9}$$

where $0 \leq T°C \leq 100$; $0 \leq D\% \leq 98$, and D is the concentration.

Table 9.41
Empirical Coefficients of (9.9)

Acid	HNO$_3$	H$_2$SO$_4$
q_1	0.5510847	0.55331383
q_2	−0.00148088	−0.002763589
q_3	0.002430177	0.002536508
q_4	$-1.5823905 \cdot 10^{-5}$	$1.3556195 \cdot 10^{-5}$
q_5	$-1.0655147 \cdot 10^{-5}$	$-1.3908097 \cdot 10^{-5}$
q_6	$-3.3085583 \cdot 10^{-5}$	$-1.0736348 \cdot 10^{-5}$
q_7	$1.6958398 \cdot 10^{-8}$	$-1.0241992 \cdot 10^{-7}$
q_8	$-4.243827 \cdot 10^{-9}$	$1.6460905 \cdot 10^{-8}$
q_9	$1.3616186 \cdot 10^{-7}$	$1.1766675 \cdot 10^{-7}$
q_{10}	$3.1740226 \cdot 10^{-8}$	$-1.0542331 \cdot 10^{-7}$

Table 9.42
Thermal Properties of Pharmaceutical Powder-Solvent Mixtures
at Room Temperature

Powder	Solvent	$W\%$	C_t (kJ/kg·K)	λ_t (W/m·K)	a_t (mm^2/s)
Paracetamol	Water	30	2.71	0.29	0.11
		50	3.09	0.42	0.14
	Ethanol	30	2.11	0.14	0.07
		50	1.99	0.17	0.09
	Acetone	30	1.98	0.16	0.08
		50	1.8	0.16	0.09
Aspirin	Water	30	2.9	0.31	0.11
		50	3.33	0.43	0.13
	Ethanol	30	2.17	0.17	0.08
		50	1.99	0.17	0.09
	Acetone	30	1.98	0.16	0.08
		50	1.8	0.16	0.09
Lactose	Water	30	3.15	0.35	0.11
		50	3.91	0.5	0.13
	Ethanol	30	2.06	0.18	0.09
		50	1.99	0.17	0.09
	Acetone	30	2.11	0.18	0.09
		50	1.86	0.16	0.09

Table 9.42 (continued)

Powder	Solvent	$W\%$	C_t (kJ/kg·K)	λ_t (W/m·K)	a_t (mm²/s)
Adipic acid	Water	30	2.89	0.25	0.09
		50	3.4	0.34	0.1
	Ethanol	30	2.0	0.16	0.08
		50	1.98	0.17	0.09
	Acetone	30	1.97	0.15	0.21
		50	1.89	0.16	0.28

From: [4].

Table 9.43
Thermal Conductivity (λ_t, W/m·K) of Some Organic Liquids

$T°C$	20	25	34.6	43.5	53.2	61.5
Methanol	0.2003	0.1977	0.1959	0.1943	0.1919	0.1988
$T°C$	16	31	61.5	76	92	103
Glycol	0.2504	0.2527	0.2581	0.2597	0.262	0.2681

From: [41].

Table 9.44
Thermal Properties (C_t, J/kg·K; ρ_t, kg/m³; λ_t, W/m·K) of Some Hydrocarbons

Liquid	Temperature	Expression	R^2
C_2H_6S	$20 \leq T°C \leq 120$	$C_t(T) = -7 \cdot 10^{-5}T^3 + 0.0218T^2 - 0.1859T + 1880$	0.9994
		$\rho_t(T) = 0.0031T^2 - 1.6304T + 881$	0.9998
		$\lambda_t(T) = -8 \cdot 10^{-7}T^2 - 0.0003T + 0.1335$	0.9983
C_3H_6O	$20 \leq T°C \leq 120$	$C_t(T) = 0.0176T^2 + 2.1098T + 2149.5$	0.9996
		$\rho_t(T) = 810 - T$	0.9999
		$\lambda_t(T) = 0.1548 - 1.522 \cdot 10^{-4}T$	
$C_2H_4Br_2$	$20 \leq T°C \leq 120$	$C_t(T) = 0.59T + 704.2$	0.9999
		$\rho_t(T) = 2238 - 2.9T$	
		$\lambda_t(T) = 0.10422 - 1.112 \cdot 10^{-4}T$	
C_6H_{14}	$20 \leq T°C \leq 160$	$C_t(T) = 0.0192T^2 + 1.5615T + 2194$	0.9973
		$\rho_t(T) = 685.29 - 1.1518T$	0.9964
		$\lambda_t(T) = 10^{-6}T^2 - 0.0005T + 0.1231$	0.9986
$C_3H_8O_3$	$20 \leq T°C \leq 160$	$C_t(T) = 4.347T + 2538.4$	0.9954
		$\rho_t(T) = -0.0012T^2 - 0.5057T + 1272.4$	0.9999
		$\lambda_t(T) = 0.2787 + 2.1286 \cdot 10^{-4}T$	
C_4H_9NO	$20 \leq T°C \leq 160$	$C_t(T) = 2.2414T + 1965.37$	0.9999
		$\rho_t(T) = 960 - T$	
		$\lambda_t(T) = -4 \cdot 10^{-8}T^3 + 10^{-5}T^2 - 0.0017T + 0.2269$	0.9983

From: [40].

The heat capacity (C_t, J/kg·K) of Bakelite varnish in temperature range $4 \le T°C \le 90$ [42] is:

$$C_t = -0.0004T^3 + 0.0587T^2 + 2.9856T - 11.659, \quad R^2 = 0.9984 \tag{9.10}$$

Table 9.45
Thermal Properties of Ethylene Glycol in Temperature Interval: $20 \le T°C \le 180$

Parameter	Units	Expression	R^2
Density	kg/m³	$\rho_t(T) = 1.1293 - 0.0008T$	0.9991
Thermal conductivity	W/m·K	$\lambda_t(T) = -2 \cdot 10^{-6}T^2 + 0.0003T + 0.2506$	0.9964
Heat capacity	J/kg·K	$C_t = 4.4483T + 2274$	0.9989
Kinematic viscosity	m²/c	$\nu_t \cdot 10^6 = 3327.2T^{-1.6176}$	0.9876

From: [42].

The thermal conductivity (λ_t, W/m·K) of water solvers of multiatomic alcohols from [43] is:

Glycerin-Water Solver

$$\lambda_t(T, D_m) = \frac{u_1 + u_2 T + u_3 T^2 + u_4 T^3 + u_5 D_m + u_6 D_m^2}{1 + u_7 T + u_8 D_m}, \quad R^2 = 0.9995 \tag{9.11}$$

Ethylene-Water Solver

$$\lambda_t^{-1}(T, D_m) = u_1 + \frac{u_2}{\ln T} + \frac{u_3}{\sqrt{T}} + \frac{u_4 \ln T}{T} + u_5 D_m^{1.5} + u_6 D_m^2$$
$$+ u_7 \sqrt{D_m} + u_8 e^{-D_m}, \quad R^2 = 0.9996 \tag{9.12}$$

Diethyl-Water Solver

$$\lambda_t^{-1}(T, D_m) = u_1 + \frac{u_2}{\ln T} + \frac{u_3}{\sqrt{T}} + u_4 T^{-1.5} + u_5 D_m + u_6 D_m^{1.5}$$
$$+ u_7 \sqrt{D_m} + u_8 e^{-D_m}, \quad R^2 = 0.9997 \tag{9.13}$$

Equations (9.11) to (9.13) are valid for $40 \leq T°C \leq 160$; $0 \leq D_m \leq 1$; here, D_m is the mole concentration of alcohol. Coefficients $u_1 \ldots u_8$ are given in Table 9.46.

Table 9.46
Empirical Coefficients of Equations (9.11) to (9.13)

u_i	Glycerin-Water Solver	Ethylene-Water Solver	Diethyl-Water Solver
u_1	0.56940289	20.541213	−89.434586
u_2	0.0020462457	−144.66498	−131.71162
u_3	$-5.465299 \cdot 10^{-6}$	151.12749	127.25071
u_4	$-1.9218368 \cdot 10^{-9}$	−18.329857	−51.687953
u_5	0.8523746	2.4210253	141.97786
u_6	0.17208439	−1.8362273	−67.311065
u_7	0.00066570843	0.60230136	−3.8573669
u_8	4.8127927	−1.9253837	106.83104

Table 9.47
Thermal Properties of Some Liquids

Substance	$T°C$	ρ_t(kg/m^3)	C_t(J/kg·K)	λ_t(W/m·K)	ν_t(m^2/c)
Machinery oil	0	899.12	1792	0.147	$4.28 \cdot 10^{-3}$
	20	888.23	1880	0.145	$0.9 \cdot 10^{-3}$
	40	876.05	1964	0.144	$0.24 \cdot 10^{-3}$
	60	864.04	2047	0.140	$0.0839 \cdot 10^{-3}$
	80	852.02	2131	0.138	$0.0375 \cdot 10^{-3}$
Ethyleneglycol	0	1130.75	2294	0.242	$5.753 \cdot 10^{-5}$
	20	1116.65	2382	0.249	$1.918 \cdot 10^{-5}$
	40	1101.43	2474	0.256	$0.869 \cdot 10^{-5}$
	60	1087.66	2562	0.260	$0.475 \cdot 10^{-5}$
	80	1077.56	2650	0.261	$0.298 \cdot 10^{-5}$
	100	1058.50	2742	0.263	$0.203 \cdot 10^{-5}$
Glycerin	0	1276.03	2261	0.282	$8.31 \cdot 10^{-3}$
	10	1270.11	2319	0.284	$3 \cdot 10^{-3}$
	20	1264.02	2386	0.286	$1.18 \cdot 10^{-3}$
	30.	1258.09	2445	0.286	$0.5 \cdot 10^{-3}$
	40	1252.01	2512	0.286	$0.22 \cdot 10^{-3}$
	50	1244.96	2583	0.287	$0.15 \cdot 10^{-3}$

Source: [44]

Dynamic viscosity (μ_t, Pa·s) and thermal conductivity (λ_t, W/m·K) of aqueous solutions of alcohol (C_2H_5OH-H_2O) from [45] are:

$$\mu_t = \frac{u_1 + u_3 T + u_7 T^2 + u_9 (\ln D)^2 + u_{11} T \ln D}{1 + u_2 T + u_4 \ln D + u_6 T^2 + u_8 (\ln D)^2 + u_{10} T \ln D}, \; R^2 = 0.9956 \qquad (9.14)$$

$$\lambda_t = \frac{u_1 + u_2 T + u_3 T^2 + u_4 D + u_5 D^2 + u_6 D^3}{1 + u_7 T + u_8 T^2 + u_9 D + u_{10} D^2}, \; R^2 = 0.997 \qquad (9.15)$$

Equations (9.14) and (9.15) are valid for $0 \le T°C \le 75$; $12.36 \le D\% \le 100$; D is the solution concentration.

Dynamic viscosity (μ_t, Pa·s) and thermal conductivity (λ_t, W/m·K) of aqueous solutions of vinegar acid ($C_2H_4O_2$) from [45] are:

$$\mu_t = \exp\left(\begin{array}{l} u_1 + u_2 T + u_3 T \ln T + u_4 D + u_5 \sqrt{D} \ln D \\ + u_6 (\ln D)^2 + u_7 D (\ln D)^{-1} + u_8 \sqrt{D} + u_9 \ln D \end{array} \right), \; R^2 = 0.997 \qquad (9.16)$$

Equation (9.16) is valid for: $0 \le T°C \le 100$; $12.36 \le D\% \le 96$.

$$\begin{aligned} \lambda_t &= u_1 + u_2 \ln T + u_3 D + u_4 (\ln T)^2 + u_5 D^2 + u_6 D \ln T + u_7 D \ln T \\ &+ u_8 (\ln T)^3 + u_9 T^2 \ln T + u_{10} D (\ln T)^2, \; R^2 = 0.999 \end{aligned} \qquad (9.17)$$

Equation (9.17) is valid for: $0 \le T°C \le 60$; $12.36 \le D\% \le 92$.
Coefficients u_i of (9.14) to (9.17) are given in Table 9.48.

Table 9.48
Empirical Coefficients of Equations (9.14) to (9.17)

u_i	C$_2$H$_5$OH - H$_2$O μ_t	λ_t	C$_2$H$_4$O$_2$ - H$_2$O μ_t	λ_t
u_1	0.094803345	0.566103	108856.407	1.84559259
u_2	0.01392092	−0.00104255	−0.0522557	−1.1219477
u_3	−0.00074603	$4.25073 \cdot 10^{-6}$	0.00668158	−0.0025363
u_4	−0.4789905	−0.0038378	1106.284402	0.3275127
u_5	−0.0132549	$-2.1228 \cdot 10^{-5}$	12546.669	$1.01967 \cdot 10^{-6}$
u_6	$7.11213 \cdot 10^{-7}$	$1.96332 \cdot 10^{-7}$	8273.396	−0.0008426
u_7	$2.08479 \cdot 10^{-6}$	−0.00470095	−15102.0379	-0.0305321
u_8	0.059416477	$2.14396 \cdot 10^{-5}$	−72284.0805	$-2.2935 \cdot 10^{-9}$
u_9	−0.0010907	0.001781598	16945.51733	$4.6891 \cdot 10^{-8}$
u_{10}	−0.00311754	$-2.1869 \cdot 10^{-5}$	—	$6.01026 \cdot 10^{-5}$
u_{11}	0.000100708	—	—	—

Additional information about μ_t (T, D) and λ_t (T, D) of aqueous solutions of methyl, isopropyl, and propyl alcohol, some acids (HNO$_3$, H$_2$SO$_4$, HCl), NaCl-H$_2$O, C$_2$H$_4$(OH)$_2$ (ethyleneglycol), C$_3$H$_5$(OH)$_3$ (glycerin), and so forth are also available in [45].

Table 9.49
Thermal Properties of Some Liquids

Liquid	Thermal conductivity (W/m·K)	Temperature Range
Benzol	$\lambda_t \approx 0.148 - 0.00024 \cdot T$	$20 \leq T°C \leq 120$
Acetone	$\lambda_t \approx 0.17 - 0.000225 \cdot T$	$20 \leq T°C \leq 120$
Alcohol	$\lambda_t \approx 0.182 - 0.0002 \cdot T$	$20 \leq T°C \leq 80$
Methyl alcohol	$\lambda_t \approx 0.212 - 0.0001166 \cdot T$	$20 \leq T°C \leq 80$
Glycerin	$\lambda_t \approx 0.28 - 0.000125 \cdot T$	$20 \leq T°C \leq 140$

From: [46].

References

[1] Metaxas, A. C., and R. J. Meredith, *Industrial Microwave Heating*, London: Peter Peregrinus Ltd., 1983.

[2] Von Hippel, A., *Dielectric Materials and Applications*, Cambridge, MA: MIT Press, 1954.

[3]   Eves, E., and V. V. Yakovlev, "Analysis of Operating Regimes of a High-Power Water Load," *Journal of Microwave Power and Electromagnetic Energy*, Vol. 37, No. 3, 2002, pp. 127–144.

[4]   Farrell, G., W. A. McMinn, and T. R. Magee, "Dielectric and Thermal Properties of Pharmaceutical Powders," *Proceedings of the 10th International Conference on Microwave and HF Heating*, Modena, Italy, 2005.

[5]   Meissner, T., and F. Wentz, "The Complex Dielectric Constant of Pure and Sea Water from Microwave Satellite Observations," *IEEE Transactions on Geoscience and Remote Sensing*, 2004, Vol. 42, No. 9, pp. 1836–1849.

[6]   Risman, P. O., "Accurate Microwave Liquid Water Data from +15°C to +95 °C and Their Use in Predictions of Dielectric Data of Frozen Foods and Tempering Processes," *Proceedings of the 40th International IMPI Symposium*, Boston, 2006, pp. 71–75.

[7]   Franks, F. (ed.), *Water. A Comprehensive Treatise, Volume 1*, New York: Plenum Press, 1971.

[8]   Kaatze, U., "Complex Permittivity of Water as a Function of Frequency and Temperature," *Journal of Chemical Engineering Data*, 1989, Vol. 34, pp. 371–374.

[9]   Komarov, V. V., and J. Tang, "Dielectric Permittivity and Loss Factor of Tap Water at 915 MHz," *Microwave and Optical Technology Letters*, 2004, Vol. 42, No. 5, pp. 419–420.

[10] Ikediala, J. N., J. D. Hansen, and J. Tang, et al., "Development of a Saline Water Immersion Technique with RF Energy as a Postharvest Treatment Against Colding Moth in Cherries," *Postharvets Biology and Technology*, 2002, Vol. 24, 25–37.

[11] Zhang, Q., T. H. Jackson, and A. Ungan, "Numerical Modeling of Microwave Induced Natural Convection," *International Journal of Heat and Mass Transfer*, Vol. 43, 2000, pp. 2141–2154.

[12] Ratanadecho, P., K. Aoki, and M. Akahori, "The Characteristics of Microwave Melting of Frozen Packed Beds Using a Rectangular Waveguide," *IEEE Transactions on Microwave Theory and Techniques*, Vol. 50, No. 6, 2002, pp. 1495–1502.

[13] Lukanin, V. N., M. G. Shatrov, and G. M. Kamfer, et al., *Thermo Engineering*, Moscow: High School, 2002 (in Russian).

[14] Wang, S., RF and Microwave Heating Group. BSE Dept. WSU (private communication), 2003.

[15] Erle, U., M. Regier, and C. Persch, et al., "Dielectric Properties of Emulsions and Suspensions: Mixture Equations and Measurement Comparisons," *Journal of Microwave Power and Electromagnetic Energy*, Vol. 35, No. 3, 2000, pp. 185–190.

[16] Boldor, D., J. Ortego, and K. A. Rusch, "An Analysis of Dielectric Properties of Synthetic Ballast Water at Frequencies Ranging from 300 to 3000 MHz," *Proceedings of the 11th International Conference on Microwave and High Frequency Heating*, Oradea, Romania, 2007, pp. 109–112.

[17] Mischra, S., V. Media, and A. Dalai, "Permittivity of Naphthenic Acid-Water Mixture," *Journal of Microwave Power and Electromagnetic Energy*, Vol. 41, No. 2, 2007, pp. 18–29.

[18] Kolomeytsev, V. A., Yu. S. Arkhangelskiy, and V. V. Babak, *Microwave Systems*, Saratov: SSTU Issue, 1999 (in Russian).

[19] Liao, X., V. G. Raghavan, and V. Meda, et al., "Dielectric Properties of Supersaturated -D-glucose Aqueous Solutions at 2450 MHz," *Journal of Microwave Power and Electromagnetic Energy*, Vol. 36, No. 3, 2001, pp. 131–138.

[20] Liao, X., G. S. V. Raghavan, and V. A. Yaylayan, "Dielectric Properties of Aqueous Solutions of -D-Glucose at 915 MHz," *Journal of Molecular Liquids*, 2002, Vol. 100, No. 3, pp. 199–205.

[21] Romanov, A. N., "Effect of the Thermodynamic Temperature on the Dielectric Characteristics of Mineral and Bound Water in the Microwave Band," *Journal of Communication Technology and Electronics*, Vol. 49, No. 1, 2004, pp. 83–87.

[22] Romanov, A. N., "Temperature Hysteresis of the Microwave Complex Permittivity of Crystalline Hydrates of Mineral Salts," *Journal of Communication Technology and Electronics*, Vol. 51, No. 1, 2007, pp. 109–110.

[23] Ahadov, Y. U., *Dielectric Properties of Pure Liquids*, Moscow: Standards Issue, 1972.

[24] Nakamura, T., Y. Nikawa, and F. Okada, "Measurement of Microwave Permittivity Using Ferrite Loaded Cavity Resonator," *Microwaves: Theory and Applications in Material Processing V. Proceedings of the 2nd World Congress on Microwave and RF Processing*, Orlando, FL, 2000, pp. 145–151.

[25] Stuchly, M. A., and S. S. Stuchly, "Dielectric Properties of Biological Substances—Tabulated," *Int. J. Microwave Power and Electromagnetic Energy*, Vol. 15, No. 1, 1980, pp. 19–26.

[26] Nikawa, Y., and M. Chino, "Phantom Models to Simulate Human Tissues in a Wide Frequency Range," *Proceedings of the International Symposium on Electromagnetic Compatibility*, Sendai, Japan, 1994, pp. 564–567.

[27] Kayser, T., M. Pauli, and W. Weisbeck, "Design of a Microwave Applicator for Nanoparticle Synthesis," *Journal of Microwave Power and Electromagnetic Energy*, Vol.4 2, No. 2, 2008, pp. 21–30.

[28] Yamashita, H., H. Kobashi, and J. Sugiyama, et al., "Measurement of Dielectric Parameters for Microwave-Assisted Chemical Processes and Its Application to Organic Synthesis," *Proceedings of the First Global Congress on Microwave Energy Applications*, August 2008, Otsu, Japan, pp. 651–652.

[29] Atwater, J. E., "Complex Dielectric Permittivities of the Ag_2O-Ag_2CO_3 System at Microwave Frequencies and Temperatures between 22°C and 189°C," *Applied Physics A–Material Sciences and Processing*, Vol. 75, 2002, pp. 555–558.

[30] Legrand, A. P., and C. Sénémaud (eds.), *Nanostructured Silicon-Based Powders and Composites*, New York: Taylor and Francis Inc., 2003.

[31]   Gorbik, P. P., V. V. Levandovskyi, and R. V. Mazurenko, et al., "Properties of Nanodimensional Silicon Dioxide Modified with Silver Iodide," *Physics and Chemistry of Solid State*, Vol. 7, No. 4, 2006, pp. 713–716.

[32]   Hsu, C., and Y.-C. Liou, "Frequency Responses of Au Nanoparticles in Polyurethane Resin," *Review of Advanced Material Sciences*, Vol. 10, 2005, pp. 325–330.

[33]   Usanov, D. A., A. V. Skripal, and A. V. Romanov, "Complex Dielectric Permittivity of Composites Based on Dielectric Matrixes with Inclusions of Carbon Nanotubes," *Proceedings of the 18th International Conference on Microwaves, Radar and Wireless Communications MIKON-2010*, Vilnus, Lithuania, June 2010, Vol. 1, pp. 94–97.

[34]   Ushakov, N. M., G. Yu. Yurkov, and L.V. Gorobinskiy, et al., "Nanocomposites on the Based Cerium Oxide Nanoparticles and Polyethylene Matrix: Syntheses and Properties," *Acta Materialia*, Vol. 56, 2008, pp. 2336–2343.

[35]   Bodling, B. A., and M. Prescott (eds.), *Heat Exchanger Design Handbook, Volume 5*, Washington, DC: Hemisphere Publishing Corp. 1983.

[36]   Poling, B. E., J. M. Prausnitz, and J. P. O'Connell, *The Properties of Gases and Liquids*, New York: McGraw-Hill, 2001.

[37]   Lax, D. (ed.), *Taschenbuch fur chemiker und physiker, Band 1*. Berlin: Springer-Verlag. 1967 (in German).

[38]   Ražnjevic, K., *Thermodynamic tabellen*, Dusseldorf, Germany: VDI-Verlag GmbH, 1977 (in German).

[39]   Hirschberg, H. G., *Handbook verfahrens-technik und anlagenbau*, Berlin: Springer, 1999 (in German).

[40]   Gallant, R. W., and C. L. Yaws, *Physical Properties of Hydrocarbons and Other Chemicals*, London: Gulf Publishing, 1995.

[41]   Venart, J. E., and C. Krishnamurthy, "The Thermal Conductivity of Organic Liquids," *Proceedings of the 7th International Conference on Thermal Conductivity*, Maryland, 1967, pp. 659–669.

[42]   Chudnovskiy, A. F., *Thermo Physical Characteristics of Disperse Materials*, Moscow: Izdatelstvo Fizikomatematicheskoy literature, 1962 (in Russian).

[43]   Dulnev, G. N., and Yu. P. Zarichnyak, *Thermal Conductivity of Mixtures and Composite Materials*, Leningrad: Energia, 1974 (in Russian).

[44]   Wong, H. Y., *Handbook of Essential Formulae and Data on Heat Transfer for Engineers*, New York: Longman Group, 1977.

[45]   Vargaftik, N. B., Handbook on Thermophysical Properties of Gases and Liquids, Moscow: Nauka, 1972 (in Russian).

[46]   Yudaev, B. N., *Technical Thermodynamics: Heat Transfer*, Moscow: Vishaya shkola, 1988 (in Russian).

About the Author

Vyacheslav V. Komarov was born in 1965 in Leninsk Town, Kzyl-Orinskaya Region, Kazakhstan, USSR (cosmodrome Baykonur). He received B.E. and M.Sc. degrees with honors in electronic engineering from Saratov Polytechnic Institute (since 1992 Saratov State Technical University (SSTU)) in 1989 and his Ph.D degree in radio physics from Saratov State University named by N.G. Chernishevskiy in 1994. In 2007, he received a doctor of technical sciences (D.Sc) degree from SSTU in antennas and microwave devices. Currently, he is a professor at the Radio Engineering Department, SSTU, Saratov, Russia. He worked as an invited research scientist in Montena EMC Company, Switzerland (1997), Chalmers University of Technology, Sweden (1999), Washington State University, WA (2003), and Karlsruhe Institute of Technology, Karlsruhe, Germany (2006, 2009). Professor Komarov is a member of the IEEE. His fields of interest include applied electromagnetics, microwave heating, thermal physics, numerical modeling, and CAD of various antenna-feeder devices and microwave components.

Index